Second Edition December 2016

Breaking Myths

Cosmic Beginning of Life

Second Edition December 2016

Spectral analysis revealed the chemical analogy between the stars and rose to the rank of certain substantial concordance of the Earth with the remotest stars of the Milky Way, even distant galaxies. In this assertion, we have the clue, how everything arises in our planet earth, including salty water in the oceans, rivers and streams around the planet. The Cosmic beginning of life, our physical environment, all we see in front of us was created as an intelligent emanation form outer space. The universe is an active mind everything moving in his vastness is use to create, and transform to the states of matter we perceive.

Gaspar Pagan

"Everything is determined, the beginning as well as the end, by forces over which we have no control. It is determined for the insect, as well as for the star. Human beings, vegetables, or cosmic dust, we all dance to a mysterious tune, intoned in the distance by an invisible piper."
― **Albert Einstein**

S H. "So, Einstein was wrong when he said, "God does not play dice." Consideration of black holes suggests, not only that God does play dice, but that he sometimes confuses us by throwing them where they can't be seen." ― **Stephen Hawking**
Hawking Is certain-

“Einstein said that if quantum mechanics were correct then the world would be crazy. Einstein was Right - the world is crazy.”
– **Daniel M. Greenberger**

"God is playing or not playing dice with the universe" The human beings play with the laws of creation, the boson of God is the instruments used to create nuclear artifacts with the same energy that God created everything; Human passions are used to destroy their creations. " Gaspar Pagan

Message received in Facebook form Stephen Hawking

https://www.facebook.com/stephenhawking?fref=n f

Stephen Hawking and Friends

https://www.facebook.com/stephenhawking?fref=nf

Stephen Hawking

If there is any group of people that I'd say have a good shot at cracking my Theory of Everything, it is certainly The Big Bang Theory. I wish them luck tonight. -SH

Gaspar Pagan

Answer: Gaspar Pagan Chevere

Spectral analysis revealed the chemical analogy between the stars and rose to the rank of certain substantial concordance of the Earth with the remotest stars of the Milky Way, even distant galaxies. In this assertion, we have the clue, how everything arises in our planet earth, including salty water in the oceans, rivers and streams around the planet. The Cosmic beginning of life, our fiscal environment, all we see in front of us was created as an intelligent emanation form outer space. The universe is an active mind everything moving in his vastness is use to create, and transform to the states of matter we perceive.

ISBN-13: 978-1537052618- ISBN-10: 1537052616

7-Introduction

In the beginning, God geometrize

Quote: "The world is a stage where everyone chooses the role is to represent" Who gave us that privilege

The magic of imagination, created in the minds of beings the loftiest real or unreal paintings, palaces castles. All that can be created in an imaginary cloud, be it can bring to reality. It is so powerful that divine legacy or sometimes not divine in being who performs, the same creates god's demons within. Through it traveling to the far reaches of the universe, or one remains trapped in a world without exit barriers being imposed. Produce dreams or nightmares in our walk through the small space of time, which gave us the freedom to choose our being for happiness or self-condemnation. Real or fictional story that has built up by the media and governments, induces us to travel according centuries of creations in the nation or territory where our soul returns as a new child, this will be our home. The incorporation of the sublime creative energy, vibrate within an inland sea in the mother's pouch, making this portal to new life awakening of consciousness that is manifest in a creature with a physically separated from his mother world, but united this by the creative emotion. To follow, who chooses the route is the human being, the result is the legacy that we face. If we abandon the creations of the others imagination, we are committed to be controlled or be slaves of emotions that do not belong us. We can commune with higher ideals to our ideas, but if we succumb to control of other human beings or deliver the freedom that God gave us, we will be their slaves.

Jesus: He dedicated his initiation and knowledge to open the intellect of human beings. He called their wisdom and teachings apply to humans since its inception. For Love of Mother, Father revealed the secrets of creation and the laws of the universe active mint; galaxies in the universe manifest the same laws, and the most sublime energies concur that the mind can grasp. Mental energy that achieves assimilate this link in the matter, may enter a communion in a real latency and understand hidden processes. A true center more perfect than modern computers, which use similar sequences operations. The millions of functions that handle these communication centers since the first human being who can grasp in our finite mind was set, the other functions that beyond our understanding and are beyond our control. Where in the universe these plans that harmonize with a physical body that evolution demonstrates superior to all that is known arise? We are a range of sums of phenomena; there is nothing to match our evolution, something that compare or challenge this type of creation. A building materials store where specific materials that will be part of our physical body are distributed.

If they knew that within every being, there an individual world tied to a free choice by the divine nature that generated us as human race. If this is understood, the freedom to create a new world within us would be the greatest legacy to humanity.

In my memory school, giving classes of farm animals. They never got much attention because every day I had to be with them where I was raised. Seeing the reality of the teachings lost time was made for me. I called more attention to a tall pine tree that flapped pushed by the wind. He was behind the classroom, where she watched out the window. In the glass, it looked like an energy flowed from its dome and rose to the 141 heights as silver strands that dissolve in the air. That experience gave me more knowledge than classroom lessons that age, a notion that I still carry in my memories as vivid as many times as long observe the phenomenon. Upon leaving I was engrossed recess lying in a small kiosk for snacks, seeing the same phenomenon, long time. It is raised with the view at this point, created in me a sense of vertigo that ran through my body to the navel. A strange feeling came over me lifting, sometimes ended in a partial dizziness. Being must be like the pin that bends the wind, but does not break. Lao Tse

Theurgist "theurgy -Work of God (Theo's- God ergo-work) (Mer- energy, Ka- soul, Ba- Body) When the control being enters the divine energies, harmonized with them. take control of your physical environment in harmony with the fluctuations. you can inadvertently enter knowledge of their true selves and their functions to overcome conditions of these legacies of humanity. the highest and refined power of God in human beings or things. the power that being can master the divine heritage of God that ordinary people call miracles. the power of the divine emanations of creation, where its laws manifest themselves. the enlightened beings who have access to that knowledge, those invisible powers who travel to this area of demonstration where the only conceives and expresses the human being.is the main actor in the Theogony of god.is the only major player in this scenario that can act according manifest these laws. Projecting stages of improvement for those without that legacy can emerge and seek a change in their lives. These attributes we do in this earthly plane is a replica of all created laws, all science, all beings and imagination. They are entities creators for divine intelligence greater or lesser degree. The gnostic knowledge, know the inner reality and perform divine laws is the only channel. The divine energy where there no name, only the essence of God. Dedicated to the fighters for the light shine and be open your intellect to reality that dwells within him, "The kingdom of heaven" That the actors on this stage made, that its mission is something transcendental.is a quality of follow the high and refined teachings, for internal growth, peace, love, gratitude, for the God of our heart. a magical power was numbed by the Fall of Man, is latent in sir and may be awakened by the grace of God, or the art of Kabbalah Van Helmond Leyden 1667

The man who discovered the potential of the kingdom within his being, has the opportunity and duty to be a servant to the causes and divine laws of our father and creator Yahweh, Elohim, Adonai. And all the names that make the human mind, Yahoo, Yokahu, Ya- HE- Vau- have. The lost word effluvium that caused the creation and the human being that is hidden in the remotest labyrinth of being. In the field called theurgist is that the journey of formation of human beings, all attributes in the creation of human consciousness and its original divine laws is broken. Consciousness is described as an attribute of human powers to realize, to mature concepts, transforming ideas, to serve, to be the repository of the miracle of life that affects the consciousness of God, because he realizes. In the same way that can debug your physical and spiritual being to rise to spiritual levels of more clarity, it can degrade your being to the baser emotions that characterize. This means that it was created as well. Which has the potential to rise to the most spiritual, and down to the crudest manifestation. For if God had the purpose of creating pure things, how did what is observed in people. It is a simple reason, the guide our inner being to overcome the opportunities you must debug their spirituality. The more the powers are disclosed in the art created by God and his laws, elements, functions, internal processes of divine creation, will be giving more power to the masses understanding of his attributes in their relationship with God. Bigoted mind of bonds, to advance the mission is create a universal mind of what brought us here to this earth for what we were created plane. The universal laws of creation to which we have direct access, since he became aware of a divine concept of creation, have been manipulated for benefit of institutions for interpretation. History teaches us the various repressive attitudes which have been imposed performance criteria in one direction. Positioning purposes interpretations acquire an institutional notion in favor of creations where in principle instead of uniting races and nations, serve to create a route shock of emotions and different beliefs. The reasons looming behind these historic stockpiles have their creators and their purposes. The route of leaders for power and domination over other cultures. No need cites examples; modern society is a mirror of these assertions. The universal laws of creation do not obey these routes temporal power. The notions of every human being are very diverse, depending on the influences to which it has been exposed, but freedom of thought leads us to the truth. The rest is simply realizing and seek inner truth within each being. The kingdom, the same freedom that we owe to God, not the slavery of being by being. We are the only ones who transmute and create the magic of its manifestations is our legacy and joy is transform and create freely.

11-The Crown Principle of creation

The process of emanations from the crown to kingdom, through the stages of creation and evolution, where divine energy be built in its interior, to declare his action, in accordance with the subject and the divine creative force. As above so below, a perfect complement to the demonstration on this earthly plane, creating an interface with the energy I think so. I must give a clearer vision for beings who have never been in contact with this medium, to see the truths of earlier civilizations, which, for reasons of historical debacles, the struggles of peoples, is the story presented to us linked to civilization. The mysteries involved in the earliest civilizations, were clouded by myths and its true meaning. It was disproved by accommodative statements, which then represented what was beneficial for government agencies and control the masses. Especially those driven under government control and idealistic currents that have been inserted in education, to ensure their interests, in separate nations. The best data found in specialized books. There's reality, which moves in the surrounding walls that story and has been buried for thousands of stories that flow into concepts disguise. They do not offer a true real front and is presented to humanity.so hidden It is that nobody notices, and those who discover are put on warning the price you pay for crossing that threshold control and protection. God loves harmony, imagination to create his delight is used. The great masters were creators of works that transcend the imagination of other beings, who paid for these creations and gods are created by possess. Only geniuses have bathed in the grace of God and His soul was exalted by their performances, beyond the material. Links with the evolution and creation are discussed generally and theories are developed far as I can think, is in relation to the studies presented on this topic. The physiological form of the evolution of beings, created kingdoms, our curiosity to know how everything works. Addressing this labyrinth is an exciting adventure, we must take all. We should not expect others build their mansions and castles within us. We are the masters and owners of our heavenly mansions.

I- The principle of creation, the insulation of the creative force emanating from the beginning, was secured in a cabin that was closed to the influence of the material forces. The thalamus is formed from the first transmembrane creation. Is isolated in a structure, where a neural center which interacts with the other glands generated in the brain developed. The awakening of this center of activity gave rise to what would be the human being at a stage where it emerged as a first transmembrane salt water. It is in the center where the heavenly paradise was encapsulated. The matrix of God that generated all that he has revealed and continue build for eons of time, in their own way and choice, with the cooperation of the laws lived has given us the power to control. Those who have degenerated the original commitment where they were given the freedom to choose and return the love who created us is the human being who distorted the way back to its original source. From some external source traveling from upper or outer scale vibrations and waves of energy projected to matter. That energy that the human mind seeks unravel their origin and qualities, which keeps us busy to seek our real origin. For the ancient thinkers and analysts of the qualities of being and matter was a challenge to the attributes of creation. The functions are studied per their internal manifestations and the results of the observations of the phenomena observed and perceived remained in teaching systems. Those created by the metaphysical and mystical systems were the bases in an introspection of the powers of internal perception. The association of the emotional effects, the fluctuation of energies and feelings that lived in different states observed. Pure science of creating what we are, bodies inhabiting a planet, we came here from somewhere as a new child and we lasted a period and disappear. Another rebirth moment arises and we wonder where all these things arise. Are we the same as we left a while ago and came back? We realize that we already have a mature, since we opened eyes to world. All kinds of issues that must relate to concepts studied exist. It is part of what we try to cover under my experiences or my way of visualizing this topic. We arrived at world with a host family heirloom, genes inherited from our parents. That chain of events recorded in the universal consciousness of things, is latent keeps things created by repetition. Evidence of repetitive cycles of existence, qualities inherent to recreate an earlier demonstration. That combination, our being is predetermined to a development of repetitive qualities in the structure of development. We realize and recognize by association most things around us in all areas, started an almost instant rapport, often recognize that many things have lived before. We sense a spiritual heritage that accompanies us, once we have training we can continue develop skills and knowledge at personal, individual level-.

14-"The Bridal Chamber"

Thalamus Eve and Adam were prisoners of outside influences. In principle paradise, then the degradation of the kingdom that was involved in the creation of being. The divine and holy baptism, where they are mixed, holy water and divine energy. The energy that assumes the structure or logos of every being and everything. The essence of God interacting with matter he created. Penetrating that legacy, to be released to respond to the legacies of the insights of spirituality; flowing from the cosmic consciousness of God, understanding of divine principles that emerge in our consciousness.

Labyrinth be it real our physical expression and carry everywhere, is our physical being that dumb shell and returns to the earth. Who can deny that reality, denies God himself and himself. We are the stage where God, the one who realizes that he manifests. forged to reality. The endocrine glands that accompany the brain system and have a specific task is determine the frequencies and shares allocated to each skein added to your nervous system component. The functions are adapted to the way the individual develops, and sensory experiences in its development.is imperative that beings take a moment and meditation the functions of your inner body, where the functions do not stop. It is as a synchronized mechanism. We abandon those impulses and draws our attention not because it is something that the eye belongs to us. At school, we are taught of voluntary and involuntary processes that border on are few who are impressed or interested in something else. The lungs breathe in a rhythmic sequence and time slots. The heart beats to a rhythm, a sequence to supply the body of their basic needs as well as the other organs. Many processes that can be interrupted by the same man for short periods of time. We can look at where we please, walking, lifting arms, or what is mechanical in us that dominate at will. But we cannot help think, grow, stop the imagination, stop breathing, stop the heart, causing the body stop creating cells; body heat, circulation, stop hearing. But without realizing all this happens like clockwork. Nothing stops while their functions are not interrupted by the cessation of all processes or atrophy of its components. Meditate on God's energies flowing through our material bodies, require a control part impresses us. We can go in depth to bring to reality in which we live. We can accommodate communication, make the real idea projected in our consciousness. Tproduce in us emotions they can be communicated to other beings. The depth of perceiving this is something mystical, which overwhelms the whole being. It is perceived as a channel where this notion is projected to our being. While the material part is, ongoing or stop, outside the sequences necessary so that the needs of human fortress. It incurs deficiencies affect the normal functioning of the body. But nobody gives us rope. We realize in those details that do not have the power to stop evolution. We can cooperate with it and let us influence to harmonize their processes, achieve greater vitality in its functions. We take food and these are added to the energy already accumulated, a supplement and create additional. Per the elements that add to our structure. The same applies to the information we process mentally, because it adheres to our thinking being as a layer or soot of what we accept as truths. We are confident that the automatic system with which we were endowed by divine nature is potential and the only one that can assume control of that machine created by God, we are.

Little account of its internal functions in our being.

Consciousness that is nothing to realize, discern an intuitive knowledge of what we are, from the observation of the combined cycle what is shown in that unfold. Another function of consciousness is that can flow through the inner matter and realize the functions.

The mere fact visualizes and educate our internal body that function to recognize the internal organs and glands, their functions and harmony with other, give us a map of where you learn about the human body. Beyond learn from their components, to visualize the internal harmony of the set of functions that are developed without our control, we would be masters of our suit A repetition of events education and maturation of accepted behavior, matured and taught those who share the same environment, family. The accumulation of data that imagination gives us evidence to situations that have not happened, but we can imagine and have a certainty of what they are. If we speak of a red house we can imagine, but do not associate a definite shape to this concept, if we wooden described not imagine a model of something or seen before. If we are given a description of all its qualities, we reached a mental picture of the house that we described. By association we accept education from other sources. For some unknown inspiration, we imagine things and events, never known, the mature in the imagination, and after a reasonable period, capture something real. In that period created internally, we can manifest on the physical plane an imaginary work that is created. From experience keep a lived reality that remains engraved in us as a film of the experiences of what is perceived externally. Realize that we can create inside and outside our physical environment, it is a dual force that is supplemented. Sometimes we go through sensory experiences that make us think, imagine an unknown world that we cannot penetrate, only a small imagine what happens. An ocean of experiences accumulates in our daily lives that we want unravel their origin. Often some internal dimension that is not common with the material life frequencies. Emotions and spiritual elevation, it like an inner channel in our consciousness that is opening elevates to a never experienced dimension. This part of communicant's energies with our extrasensory experiences are cataloged in a cosmic or universal mind, from which everything emanates from the divine mind where everything is contained. Attune to these frequencies it gives us a certain awareness functions and internal attributes that allow us perceive that dimension out of this plane.

Lower forces in the thalamus and our endocrine system. Display our inner being begins communication across networks emerging from its center in communication terminals or nerves in the central nervous system. This their look is an extension of neurons systems graduating constant combination of materials to maintain a flow of energy regeneration or new construction sectors actuation pulse. The central nervous system is the communication channel and signals which is adjacent to the skin, which serves a tuning fork of reflections and exchanges of energy, interior materials, and traveling to impregnate or harmonized with internal processes. We are a major magnet energy that constantly combine to complement the body aggregate materials added inside, through the consumption of food materials. Aggregates energies of aspiration, liquids, and compounds which entered the system, the aggregation energy adhere to invade inside our different external sources that combine what is being. A network of veins, capillaries, arteries valves and communication tunnels encircles our internal system. Through it all building materials that are the body has acquired for construction and aggregate materials transported. The secret that many cannot access that process is like the body, the energy of creation separates each component for their individual qualities and directed by the internal distribution network. Precisely the place at the amount, with an internal intelligence of what is what and what their function. Before adding the function to be replaced, and when get to exact spot, start their functions without instructions. A channel feeling for each distinct function. Each function has a sphere of different feel and separated from the others, to meet at a given point that feeling is united and added to the other, forming a circuit event and hence the cycle of many others that are added to the primary sensation. But we know that is not a small center more complex than a computer operation, is in the center of the head, called the thalamus. It is mentioned in the books of science and endocrinology. Its true function is a myth for most those who do not enter higher or scientific studies. I have not familiar with earlier theories, or Darwin, I've only heard in high school. What is, is something intuitive readings partly paid by impartial, so far and acquired knowledge in esoteric subjects. Intuition is a channel which moves in the subconscious memory, is a feature I Call the communication channel of the material with the spiritual. The border where the divine forces meet with the physical. Being that senses and realize the essence of God, achieved through these sources of internal communication where he is revealed.

The other part is a little maturation of how control the flow of internal energy by concentrating attention on specific areas, setting mindfulness and visualizing the internal functions. If put attention to the muscles of our body, we will realize how these automatically react to the reduced need to apply its functions to a bodily reaction to perform any movement we want action. The reaction is immediate and the reorganization of millions of muscle fibers is formed into the exact precision to action to be carried out. Another way to train the internal system in which we can influence, is concentrating additional energy, do repetitions of functions determined outside of the automatic reactions learned by the body system. We can create our system control certain parts of the cycles depending on some areas, I know the processes that we must apply. This was a wisdom of many mystics who managed to perform functions beyond what people call miracles. It directing attention by means of visualizing the internal processes, while a portion of breathing air that adds energy than is present in the breath, so that the material body pays more concentrated component, the construction process is applied inside. It's like a cooperative process of maturation and aggregation of a premeditated notion wrapped in a natural process. It is a process learned from ancient wisdom schools, to harmonize the inner energies of the body. Within this exercise an extra notion of something filtered. only known mystics prepared for millennia. An inner notion of a spiritual force superior to any matter which eventually processed. The notion of perceiving God.

The magic transformation of the inner the steps are internalized to provide the mind of those higher functions. The primary impulses arise from the need to make them the environment and its components. That first egg creation binds of proteins that cells begin to produce by the thalamus; (the first egg of creation) generates an internal process of inner magnet attraction or what we call consciousness or individual intelligence. It becomes necessary impulses or law of attraction to other elements that provide electrons and complement their development. Notion that quality of matter and its components, is both divine intelligence that emanated toward creating and isolated herself in a cabin where he started control and stages of creating your own being. In a voluntary process, they can influence the internal currents, emerging from neurons to travel the complex nervous system clothing fully functioning within. At first this was a reality for people, with its attributes manifest itself as it is. They are legacies that do not recognize today because processes have become in automatic copies. Its effects do not occur in random changes. The memory created for these functions are already aware and distinguished one another. The constant influence of gravity on our physical being, unlink most of our accumulated during our life on earth learning experiences. Our faculties are a large cluster of conditions by the force of attraction. If we got rid of this, the vision of existence would have a change of perception. Countless items that come to organize this process in stages and adding time as needs be created. In each entity acting, must take different routes, structures vary per the aggregations that manipulates, in each subject created their function is different. An energy that is wrapped in magnetic fields and harmonizes all its laws.

His appearance was in the first bodies of salt water, the first reactions to the energies released. The first cells created a kind of magnetic fence were detected around other energies that were not related with theirs, to reject what was not harmonious with their structures. This addition was creating various reactions, due to its interaction with other creatures that developed in a common environment. Eventually a kind of magnetic field responds to what we call aura was created. This is an energy field that responds to the energies that act within the functions carried out internally. It is endowed with qualities to detect other energies, at a distance, thus protection areas and rejection of what approaches are activated. Was so necessary in the development, many species develop rejection systems, most sophisticated you can imagine being. Electric shock, camouflages, extraordinary sounds, or direct attack, with attachments believe that the situation in their structures for defense. If Analyze their structures were not born with these elements, these were created by the need to overcome conditions of their environment. This is the nature we see in another species. The magic of transforming matter into structures control laws is a mystical and divine legacy of our creator. Since assuming the functions of laws and their new creations, he began to interact and magic of transmutation of matter and divine forces within the self. The power we have by inheritance is a divine legacy, we are the possessors of divine mantle of creation CERN.

21- Boson the God Particle Discovered

Laboratory in the statements of the year 2012 is described in the news that it has finally found the gender particle creation from an infinitely small particle or particle of what all was generated. Archimedes possessed mastery, intellectual strength and mathematics to show 250 BC, the time and space concept. Egyptians also took part in the accumulation of this knowledge. Interior lighting infinity is the same as the God particle. If we cannot even conceive within us as his creation, it is nothing but demonstrate the energy that moves us. The way the mind or subconscious are impressed by this energy is equal to the reality of cosmic energy. If the real awareness of this dimension, material evidence, all that shows is that we are the consequence and part of it. If we are not able to realize that we can describe what is or discover its last particle, at least it is more than a physical impression of what it is. The depth of the human soul could transcend this dimension.

Summary-Gaspar Pagan Geometry was the first expression of the divine mind in his first effluvium where the spark of creation originated. - Kepler 1619 The order of a figure and harmony of a number, evoke all things Giordano Bruno Johannes Kepler -1591 God Geometrically The quintessence Plato was an Egyptian teaching where the world is formed in its tiniest particles by triangles grouped into five elements or quintessence. Kepler corrected this proposal Plato and enunciated that God entered exactly five bodies between the distances of the orbits of the planets. With this argumentation discovery "I've stolen from the Egyptians the five vessels of gold, to erect my God a sanctuary away from the borders of Egypt. The five intertwined by the planetary forces that make up the Octahedron figures." Among them is the divine energy that only they handle." Gaspar Pagan Archimedes had a great mentality and demonstrated equations the same principle in geometry. in any case, mental alchemy behind the Kabbalah, gives the idea of the fall of the higher bodies, was triggered by the fall of Adam. the divine degeneration to be created. Paulus Ricius 1516. the man has not stopped his hunger to simplify or theorize God. scholars who meditated before us, obtained a deep understanding of the cause primary. current beings walk floating on the diatribe of who or what best describes God. If the clear majority of humanity accepts that it cannot understand and communicate internal qualities, superior in the cosmos envelops its functions that same cosmos is for loved the God of their consciences. Scientists reduce these functions, the microcosm, atoms, electrons and protons, or particles which document how the common being only sees the macrocosm of its manifestations, and these scientific deities and cosmic forces the catalogs of God. Attend Will find that his concept would realize that we are the only ones who do, if human beings did not exist, said, I God would not be a reality. God is the mind that is done. The primary energy manifested in humans, but not the human being itself. Man, is only a channel of expression of the primary force that encompasses entire cosmic universe in a real manifestation of it. A shadow of what began volcanoes that constantly interact with matter, created some lakes where the basic elements that clashed with the mass of matter that was the beginning of planet earth concentrated. To his credit already it existed where material forming was expelled until reacted to the laws of attraction and repulsion and fixed in an orbit in the solar system. The mere fact of occupying an orbit in a complex of galaxies in the universe, gave us that quality of being what it is. Which, in this small cluster of materials and energy, have concentrated in this atmosphere, it is the demonstration that our senses perform no matter the stage of evolution where we positioned. Sequences of matter and all its components from the simplest to the most complicated, merged in the states we all see every day. What mysteries within these is the challenge of human intelligence. From there its core began to form compounds with hot lava eruptions, creating water condensed for long periods of time. The layers were cooled and the holes created by the rock that was not sealed were provided as strainers matter liquid where are concentrated. minerals diluted in the mud resulting from these leaks and mixes with sediments Over the time the first simple bodies, demonstrated in the breeding grounds that accumulated and recreated the first cells. With reasons these reactions and were dragged to the streams of water that accumulated in the basins left by the formation of the rocks cooled in the seismic activity of the earth where large amounts of energy were released into outer space. In that seabed, they are starting the first cells to form the matter that enveloped. That formation is the first emerging characteristics, given to the reaction that arose in individual entities, as the combination of matter and energy to which they were exposed. In a game of strength and energy, the internal components of matter are manifested in a reverberator of atoms, energies of primordial organization. Layers of material being accumulated and organized into something with its own characteristics, such as bacteria that gave rise to fungi and yeasts with chlorophyll to absorb sunlight. The energies emanating from space attracted by these reactions were concentrated first on the water (H_2O) - (O_2) -for was the source of attraction and combination, which lent to the elements of the energy will be combined into materials of different subject categories. The interactions of solar energy, the moon, the emanations of the planets, cosmic energy of galaxies spread their attributes to what was formed as the planet earth. Magnetic forces of planets that keep this in an orbit influence the nucleus of cells created by establishing a dynamic identical to that of planets rotation. Centrifugal forces of attraction to the bodies. The continuous change of forms is due to these influences of external forces on the atoms of the thing manifested. Within a layer of matter of related electrons, a fissure created by the same reaction of these energies is opened. Its structure regroups in layers, with varying degrees of congregation and attract other sources of energy that balances its structures, in the same way that destabilizes, this principle also seeks force rest. The process of creating layers to isolate processes makes them, exteriors and interiors. The dynamics of this sequence creates an interior cabin, where energy is blocked between two walls. Rebounding of the energies released from this combination, strangled in these denser layers do function triangular spaces, where matter begins to create extensions thereof. Hence a natural reaction is that the first cell created gathered his forces in the center of his being or vibration control, it vibrated Along with light-energy focused on her. He moved by impulses of attraction to the source as projected by its very nature, bounced reflex (a reflex the same as a mirror) doubled radiation of the same quality. The shock waves produce other quantities entering the game harmony of what is interacting, some of that energy is released.

They leave their orbits to be attracted to other materials found in energy imbalance and projected outward. Those that fail to be caught by ripples or other nuclear orbits for some behavior are drawn inward form a triangle. That combination was trapped in the walls of the undulations of vibration in a circle, where energy is concentrated and radiate and established reflection. The first notion of circular energy affected being, the creative energy takes shape in what would be a new entity created. This comes to know its source to react to the dimensions of the internal movement, their ability to react to movements and discharges; They trained the immediate and continuous reactions. With the concentration of primary material, in this case sodium, potassium, oxygen, hydrogen, carbon, that were present in the salt water by concentration. Proteins that give rise to the first membrane at the current position of the earth, or as an inherited from its previous orbit where detached attribute are created. The cluster of reactions gave the base to the first membrane which created a specific matter, proteins, so their reaction to remain actively concentrate on it, proteins gave support to survive. The primary cell formation was androgynous energy of two polarities which were separated by the same laws of reaction, each adopted their qualities of reacting with other, the attraction one another and strength was manifested in light waves; They had rejected each other for the primary chaos; which gives rise to the aligned polarities to their qualities chemical activated by the flow of magnetism. Penetrated the first membrane formed matter, in its movement, was repelled by the repulsion of the other energy, separated by space arises attraction between the nuclei created. In this fusion of attractive forces first reaction than it would be the first cell of the matter which gave basis of the two lobes of the thalamus is established. In this case I am oriented to describe what is the principle of human being, not avoiding the creation of other species and kingdoms that occurred at the same time, but with different vibratory rate emanations acting matter and different conditions. Internal emanations recreated the first matrix of the first membrane formed. The combination of these emanations was mixed and fruitful; Thus, he was born the strength triune love, interact, harmony and involution (A miniature universe, the central force). Of that cosmic harmony, he was born the first emanation of a cosmic symphony that encompasses the phosphorylation of white light into seven basic colors. Comes the first harmonic scale, which gave basis to the projection of the elements in the cosmic space. The verb the first law completed the triangle of the first divine action in the matter, being mother father with two human powers) - created the missing dimension. The divine manifestation of primary energy interacting with the vibrations of the infinite mind, (the energy generated from the primary source of vibration, which reached this distance created a unique quality in the universe). Which became aware of himself, in his first conformation. Emerged the material dimensions of being, the dimension that shaped a reality that simultaneously reflect his presence, arises. the creative perception of God

Where not exist before this notion of existence, it was spewed into the material manifestation where he began interacting divinity and being created, high, low; light and darkness; right left; notions of divisions graduated energies to give the first active creative energy to mind. To express itself, it degraded to point of becoming known and become conscious of itself, but in a dimension never manifested. A dimension in which trained herself to flow with the creative freedom. Ion permeability both sides induces a transient change, generating an electrical impulse. An arc shaped meeting cloak that enveloped the matter, in reactions of attraction and repulsion forces, creating a balance, a medium, the annulment of the opposing forces, the lobes of the thalamus. This creative fusion emerges from the alignment of different forces emanations different distances. Being in the same orbit was added creating a phenomenon first manifested. Conjugated double energy creating a new dimension polarization interlocking lobes without shattering the thalamus each other. Something the imagination of ancient only knew because used in their resurrection and meditation exercises. That polarity triune force is created. A fourth force comes in from the outside, to influence other functions, in periods distant of time and strength of maturity. spiritual While these energies interact within the brain that engulfed entire structure in a corpus callosum or skull, to regulate energy rates would enter his cabin. Your endocrine and developing body. That way would only spend the amount of frequency with which you will interact harmoniously. In addition, the thalamus creates its interior with the same principles which regulates the energy and vibrational frequencies, which filters for adjusting the quantity and quality of energy that combined in its center. Axons and dendrites that abet also have their sequences downloads, like dendrites and terminal development to channel these frequencies.

Studio with tests on the atomic interact. Basis of quantum theory: "With the new century born the new doctrine, the December 14, 1900 suggests Max Planck (1858-1947) the innovative idea of considering the radiant emission as a batch process that is carried out by isolated elements energy, holders of a certain magnitude. This element, as is proportional to the frequency of lightning, being the factor of proportionality a universal constant of nature, the famous constant h which was later immortalize the name of its discoverer. So, the energy of a quantum is given by the formula. the lucidity of this thought clarified hit the enigma of black body radiation immediately explaining the variation of the bell curve, whose caprices had puzzled researchers. Such success was not but the first achievement of the new theory. Assuming the fertile germ Planck most unprecedented and wonderful ideas that should transform beyond recognition, the image of the physical world was hiding. " "In that extra dimension not altered by this energy field, it is that divinity outside the influence of other materials is recreated. This gives the be developed after the thalamus gland, a unique quality to become aware of all its attributes. A brain than other animals created qualities. A condition that allowed him shed its primary energy and is raised in harmony with the same creative energy that generated, sharing, (open a channel of attraction, mixed) with the source from which emanated your condition. Communion as can be called two pure energies for periods of sequences and necessity can attract to each other and merge into a single universe of pairings consciences than they are. "The notion of life before death, emanations and emotions, passions that occur in a newly created awareness, enjoy the first emanations of the cosmos in the universe. What must manifest, men women contained in one egg and reciprocate their emanation inside. As a radar that returns the same signal that is received with the usual reflection of its counterpart that come together in harmony to create. Acting in the consciousness of being the mystical elevation could enter for short periods of time in contact with the divine energy, exaltation only channel for the cosmic, God. Just imagine that this gift can be reached by humans or in specific periods of human history has been a knowledge that has not been lost, that is affordable to anyone that matches him, I imagine it must be a gift for seekers of divine wisdom. It is something that many aspire to own. In this analogue world gathering forces and created by developments in the primary force, interact and needs follow a pattern of evolution towards demonstrations already created. As a scroll or duplicate of the first, but in a far from its original center dimension. Where the characteristics differ in light waves and refraction. The imbalance qualities of attraction and repulsion, an already expressed need passion to continue projecting new and unknown things. A new heritage of primary energy that was attractive, brought a harmony of his father. As the womb of a mother to give birth a new creature and energy of excitement, seduced her to follow a pattern. As it manifested in a new flower and its fragrance, with a superior will not owned by anything, just being created the image of divine energy.

Awakening

Without realizing it, I opened the door to the mystical development ancient wisdom, which involves the aspects that awoke in me being asleep for millennia. In my conscience, I was harmonized with them in a tear of energy, which joined the abstract world, not border with being. Consciousness with the world of physical expression, in a kind of cosmic initiation, a sense that opened the channel where many mystics have been touched with the link that lies beyond human reality. Contact with that sublime creative force that reveals the other side of creating the site where intersects the other side of light. Some ancient mystery school for initiates this rite disclosed or drawn from the universal mind was saved. The awakening of the subconscious that covers everything like lightning going off and lost in the remotest corner of memory. A torrent of blocked energy, moment the rational force of matter transports me to world of irrationality where the spiritual is confused with reality and gives a vague notion of the inner worlds of creation. Since it is becoming aware of God's Creation and the energies that give rise to fluctuations of these routes covering the development of the universal. Worlds that have been penetrated by many search engines and inspired by our father loved God most high creator, logos infinite Universe so called because it is universal to created beings and their poor understanding. The inspiration of so many mystics and avatars were surprised to cross that threshold, entering a stage lighting, like many before him. Legacies are, the reaction of the beings who deify mortals father couriers. God created the universe, the material realms and ultimately being, who gave their greatest spiritual riches. The biggest secret is that was their creation and deposit at its greatest treasure, the kingdom of heaven, the kingdom of the inner light, a miniature universe like the grand universe. Where all laws are contained as in the divine mind. The lost word, the word of creation, the Grail (of creation) that contains the real name of father-mother God and all the names of creation, not being before becoming real. This link lost and unfocused, mimicked in the consciousness of being, where it was first recorded in the thalamus, the matrix of God. That begat itself from the first emanation of evolution, where the father-mother-which his name cannot be pronounced, which comes from the light of the heights, the father of silence was recreated; light of the word, divine powers radiation to the earth auto engender impossible to be grasped by where the light, humanly cannot be created by the light, being cannot decipher the vibration put in moving the first law. Where duality created the vacuum of the demonstrations, it was motion and rest, to divide grew and multiplied, concentrating the forces of opposites in the center. Being dual, father-mother, and concentrate their forces at its center, the son of motion and rest-law of the triangle that dominates human manifestation of divine arises. Being that arises is the only one who realizes God and realizes his being that interact fades from his first essential but does not lose its quality creative-love harmony to the new divine dimension as a binge creating divinity. The mind of the creator is where content was from the beginning, emanon toward being created by love. Remains asleep and buried under thousands of layers of creations and under free will, the power of freedom to choose, sometimes erratic notions. Human consciousness has diverted connection with the divine and disconnected with the ethereal. The immediate material seized his conscience, lack of harmony with the fourth energy that manifests. It is the work of architect divine on earth, the cornerstone, open a channel of attraction to the superior force, the part that gives that legacy to be, it is part of their way to it. The cornerstone architects rejected, because they had no true knowledge and were dazzled by the agate stone, nature. His intellect was diverted from the primary source, and lost their orientation to the present. Secret of Secrets, the silence of God, to those first beings, builders of temples, heirs of Adam's mind.

This fourth force and lack of use of intuition, was buried in the remotest corner of memory. We have the power to wake, but the real paths, the right lighting to attract this force to us in its original state, have cloudy and has become a dark shadow of our consciousness. The secret that dominated the alchemists who tried to enunciate those principles to the more advanced beings. Each day has been removed by the material of the same being's ego, misunderstanding, lack of orientation toward what we truly in reality. The union of opposites energy, pure light and darkness, in the bridal chamber, the sacred place that is contained in being, the seed of creation. The confusion of the divine nature with the materialistic reactions to slumbers and drunkenness. Being must return to the light and soon be on the right track.is time to abandon old concepts that contribute nothing to the spiritual elevation of humanity. The truth ancient wisdom is pure, like it or not the detractors of these notions of wisdom of the early mystical science. They were the right approach and left their legacies embodied in the truth that man is master of his own free way to parent. It is the divine law and the relationship of being with the creator is direct and personal, of every being to be in different stages of evolution and understanding of its internal attributes. The contact and exercise that power in line with that abandoned for lack of guidance, where we went away to attribute. accept get that extra gift another human being, the gods created

The creation statements for the first philosophers assumed to a creator God through the word acted in the matter and in humans. Demonstrations combining the light of the heights with the matter formed in the realm of earth, which is the only one who could be conceptualizing of the phenomena that manifest themselves in the projection towards the earth and created kingdoms. The main reason the study is the human being and how it was created and how the divine master of laws works. Phenomena I describe in this system internalize the divine emanation of God in the labyrinths of human consciousness and its attributes that include their personal logos and collective. Descend from the spheres of light being encouraged to matter that connects the crown with the divine realm. After a phase of manifestation and ripening investment process where light energy returns to kingdom for regeneration or purification and enter a stage of suspense until the time of the law is fulfilled, and is Capable flow from return a new creature at birth. I give here an example that describes the inner journey and everything that happens when emissions interact with the conscious mind.

My initiation Living the freedom to feel the divine energies of creation are, a path expression that occupy spaces within the functioning of beings and that they can participate in that process. Maturing an internal training and make real that divine spiritual gift within us without anyone, oblivious to the dual process of being and God, be interfered. That was the way Jesus and the early settlers interpreted their relationship with God. It follows from the Hebrews, Jews, the cultures of the region where the Old Testament and part of the story emerges not changed. It is mentioned in the ancient texts bequeathed to those who took control of the institution to collect ancient writings, guarded by followers of Jesus. The stages of teaching, trained all who joined the Essenes and other schools of wisdom, knowledge and maturity through a development and preparation were offered the opportunity to belong to the inner circles of the brotherhoods. The method was maintaining the purity of the teachings, which were available for all seekers. I went down to the inner shells of man, my own being with all my attributes. I untied the evidence of a soul in decline, I saw exceeded as the light began to shine properly, its essence is away from the material presence. Inside lit each day more and outshone his own being while returning to a state where He stopped the advance, diverted for reasons that marred my understanding. He knew that road and followed, and returning to the paths forking before me. I had no idea at first to where the tide turn, the same as I would have to follow to know the gift of the elect, the wisdom of the adepts interior lighting, which my being guided my desire to learn and know. Was just another deluded being, seeking the path, the path, then jump emotions defeat, internal inclinations doubt the lack of certainty of what is sought, clouds understanding. If you look in intangible legacies, it is sometimes lost the trail, although it is believed safe. Walking on crystalline waters without seeing the veil that sustains the soul. Feel the hunger of knowing lost and want more, look around as everything crumbles before the futility of the search. Be on the road, the right way and moment the glass hidden in the path disappears under the feet and soul sinks. Return to beginning and is a routine without return, then re-take the path flee and mysteries hiding in the walls of time, created ghosts by offset from the true sense, have ended in the same crossroads What maze of shadows hovering over the search and suddenly a glimpse of royalty looks and the soul rises, knows the reality, begins the darker path that humanity has traveled. But not in its material spiritual attributes but his poor soul that since he was kidding and released his first love I created was distorted itself of light, its original divine heritage and the purity of his kingdom. The more sought the light that genre moved away more of its creator, outside the effluvia that left flow, float and out like an inflated balloon of his own being. Ignorance of the true path was diverted by those who introduced a remote practice of original knowledge. Tyranny cruelty led to deviate from what should be the rebirth of humanity. Those who had the original knowledge not paid to those contrived creations, and no one could understand the message of the Master, who knew not bequeathed to others (the Nicolaitans), because its purpose was that his people not succumb to power of empire. These were mystical traditions of the Essenes and Egyptians, Hebrews heirs of the mystics of India, where the original knowledge was developed. only It was affordable to those tested in their arts. To know the divine essence was necessary pass a preparation before being admitted to the preparation and demonstrate at least one year the mettle to be worthy.

The Legacy and arose in the land division of kingdoms and creating what would be the father and creator emanation of the universe was expanding in all directions. The essence of creation that travels through a tunnel of light, up to cosmic portals where once you cross will not return until the train regeneration to inhabit a new body, called by human death; by mystics, transition. This return to life after his Reincarnation. Energy light emanates egg of creation after a period of regeneration, will come to light, his breath came into the cosmic soul. It is housed in the new being that cry to their parents to get back into a material body. In the early dawn of awakening manages take control of your new cockpit and start your own domain and will join the new material the spiritual attributes that accompany this move of evolution.

34- The secret energy

The energy return of the father, incarnation of his light and be through the union of opposites and again return to a new creature where he joined part of its essence with the new dwelling of light, so that a soul is a reality in this earthly plane. The same law works in the other kingdoms created by the highest divine being. Except one condition that differentiates the human being as God Creation.

Page 11 (The Amazing Human Body) digest- Reader ISBN-84-88746-23-7 centuries that free the human being, a mere detail of creation, the quality female, hormones, estrogen determines the maturation of the sexes or the maturation of their bodies, sex. It is an explosion in information stored by its laws. And for me the first genuine and reliable detail of how all started.

The female clitoris is developed in the same area of the penis in males. The endocrine system is the brain that stimulates the ovaries to release estrogen. These female hormones are responsible for the maturation of sex organs. I take days to meditate on this new phase of creation. Thus, in one body the sexes alternate in each incarnation. Which it assumes that being originated as an entity occupying one body and then emanate all the features of the evolution of alternative. The perception of a space that spirals climbing and travel back to establish contact with the cosmic forces. In the first stage of creation manifested itself in salt water, the feminine force emanated first and top scale supplement to the law of creation divine energy manifested father and the egg of creation was formed. The emergence of life in being as known for all eternity until the last effluvium that being can grasp, because he is the bearer of his kingdom. Being the beings that have been created to give a light of divine nature, the ego of our way to control and ignore the divine nature of God in our own being, because has been diverted by other beings who try to preserve this knowledge. They have not had in the way give out something that lasts and is consistent with the universal laws. Those who project an idea of it known, have embarked on systems not let this flow freely into the minds, as they blocked the legacy to remain ignorant beings. Just be affordable to the chosen and prepared to handle this knowledge. With these original spiritual gifts, which have managed preserve a cache of secret formulas and arcane than it should be something routine methods. As life progresses more, wrapped in layers of protection something that should be natural to people because is the bearer of that heritage the kingdom in his being. It is the human being, who inherits these functions in his being, you should have access to this knowledge and become aware of that gift, that the creator has placed in him. Others are elitist systems to use one created to manage an asset acquired adepts who must maintain an ancient tradition that sometimes verges on fanaticism version, because that was imposed, how to make it known. Long before this period of obscurantism and domination, the advances of science and health towered as some advanced branches of humanity. Although they were involved in an imaginary world of ancient gods and mental creations that have come down to us in books and stories. The contemplation of forces disproportionate between men and imaginary gods, to give an explanation to what is imagined. The deification of beings for their qualities, making the sacrifice of his own life, giving his soul to God in sacrifice, this was the ritual of the gods and ancient deities. Examples daily living and not by scholars but by ordinary beings who not mastered the sciences. The nature of many human beings who have gone through this personal experience and have returned to life, can give an account of this astral journey resurrection, where they did not cross the portal and had the opportunity to return to the same material body. No way this trip is a radical change; It is just a bridge experience conscious and the subconscious. An experience that human beings are endowed these attributes are part of their divine inheritance.

Reincarnation only see and feel its expression in this earthly plane. The we conducted as a matter of routine in our daily lives. Are few who are passionate search for items that enter the game creation. Many mystics penetrated in search of principles, or causes of our manifestation on this earthly plane. Their findings fertile ground for the human being aware of their own existential reality. Civilizations that have contributed to this knowledge have been studied, the greatest contribution to humanity has been lost, wars, destruction of their legacies, the removal of the original to impose something fantastic, which benefits to its creators. What has come down to us, is the collection and disclosure of one part of the story; the other was hidden or destroyed by historical debacles. The other keys were buried under a dark knowledge and only a few have managed penetrate that wisdom. Many concepts have been eradicated from public knowledge, for trying to document these aspects of universal laws, their creations lose the value and forth doctrines are evicted from their power by creating a vacuum in their statements. A vacuum opaque the path of light. An empty in content not being motivated to raise their intellect to perceive the God who created us. Content filtering human emotions to focus on a figure that existed at one time, so the notion of God to created beings is limited and not the divinity of father of creation. The amuse or divert emanations of souls who live on the physical plane makes your inner divinity emotions are focused on a concept that is not correct. The degradation of perception is diverted to messianic concepts, where an income is perceived by describing a message of what God is believed. No respect for the principles they proclaim, abuse of power, the misery of the poor, not sympathizes. On the contrary, use their resources extracted from the plates of the need to speculate in politics, without fear of offending the god who claim they are whited sepulchers not flinch before the violation of laws. They are the less they feel elevated to do charity and are afraid of God. Education systems in space is not provided to alert the interest of finding spiritual universal laws, they pay to our existence. Maintain an intelligence that varies in degrees of appreciation of imagination and preparation and spiritual heritage which returns in a new soul, a child. Globally, maintain relationships with those who are separated from us by material and intellectual barriers. Are few who have managed overcome the fear of persecution and threats of statements that were imposed on human beings under terms of controls to protect assets. Subconscious culture still survives in that stage of civilization lived. Now, with advances in global communications, everything is at hand for anyone; the world has shrunk; there are no barriers to stop knowledge. We must take advantage that freedom gap that has opened for our advancement is straight and real. Those who hold the power of education must provide a system for creating awareness in the educational part of this knowledge.

38- Original legacies

It was a sprout suddenly, and after a process of evolution which evolved over millennia attributes that each he matured physically and internally. Internal and emotional awareness enables us to build features that give us a perception of superior forces, which dominate the universe it belongs our planet and the humans who inhabit, which, therefore, are the only sentient beings from this cosmic stage. So far shared without knowing that apart from everything that is manifested on our planet, known or unknown. Out of that figure is a mirror that we someday get to know. From the first settlers, we have experienced the most diverse feelings and coined within us as something greater than ourselves, we venerate because many of these emotions are spiritual. By spirituality, the human being does not a clear concept. We assume that is the belief in something divine, an attribute of God or legacy that gives us higher emotions, which we have access. An interior recollection, reveals something higher than what daily thought, feeling emotions, admire nature, seeing the face of a child, the closer to something that is proclaimed as divine, was a face a relic. These are internal reactions from those projected values, which are graded spirituality. Try to explain personally, which is a legacy that grows with life, matured by evolution and application of new sources of imagination to knowledge. Each being contributes, as a separate entity to the great universal mind. The subtlety of these is at a level that we perform in our interior and daily living, but our vocabulary cannot describe in words those inner realities. Philosophers, physicists and scientists have expressed about giving the most diverse explanations. The causes operating in the creation not agree in their conceptions, that took root in our minds since we studied the concept of our own existence. Scientists continue search for the origin demonstrate that quality of life. Oceans of ink accumulated in libraries from stones to today technology, where stop time at different stages of the internal events of our imagination and capture in some means, communication. Science, genetics, astrophysics and other branches of knowledge and catalog knowledge of the cosmic and atomic and its functions have advance information centuries; to achieve spiritual and physical catalog of what comes from cosmic matter. The magic of the mind over matter and the desire to create by the same free choice. It gives us the freedom to divine wisdom enter direct control of certain powers of the imagination, joining one point to another and reflect its essence in all directions to create a new underworld; I was not in the mind, but is possible in the divine mind has no limitations. The mere fact comes realize in relation to the thrill of capturing the attributes that the divine mind has access. With this simple step is that the small door of the universe is revealed in the soul of the being who seeks uniformity of all laws. Perhaps sixty years ago, it would have been a great job overcome these barriers. Since the Moon first set foot on the man jumped in his development has not stopped. We are overcome the concepts that accumulated over millennia by the best minds endowed us with basic knowledge to educate and continue the accumulation of data in our respective inner worlds. Colleges and universities still use these legacies of antiquity as a basis of phenomena which left us. The most significant change was diverted mental stagnation of religion with science. Put life in a new perspective, overcome concepts accepted as truths. Knowledge must be a tribute to those mentalities. The world is witness that has awakened a new era of knowledge and maturation of human intelligence based on concepts that challenge the most established schools, although still ripen a world of concepts, not lose that link, that door which gives us the freedom to know our past. Scientists who shaped the bases in the cognitive, as an abstract universe of knowledge and projection of human thought to an extension of the creative energy that exists and to create and great mathematicians. Archimedes was one of those gifted teachers few years before Jesus, their treaties are being rescued by scholars who, although have the original strive to give mankind some hope, in 2005 recover data from your started C treated, disappeared from the library of Constantinople. The code Archimedes manuscript that could have changed the course of history of science Reviel Netz-William Noel (ISBN- 978-950-04-2926-9)

Who does not suit you that this is discovered? At great defenders of truth, thousands of names were contained in the libraries of Alexandria, no one knows the destination of Constantinople that was treated without mercy. world Center that will provide humanity of ancient wisdom and an advance without many obstacles that must overcome. they are ready laboratories and advanced minds of humanity waiting that great outcome of the hidden history, that being drink from their original sources. this can be overcome ignorance in which it has maintained for centuries. the way out of material concepts that Allegre us to a universality of all attributes of matter and universe energy. to that goal should be directed the intellectual forces of universal youth to bring together the real concepts for lasting world peace. The day that our planet cannot accommodate more beings, our youth should have ready the way to colonize the universe.

Evolution Tests Crown (Keter) The thalamus in the center of the human head. Esther Gómez Rosón this Doctor's contribution to science and scientific future: In a simple way, understandable gave original internal processes, human contribution that still exist in our original system. In the scientific study published by Dr. Esther Gómez Rosón in 2005, you can see perfectly the center of the body. Each one develops an individual personality, which is linked to the functions of the thalamus, associated with them endocrine and system. hormonal It is the field of action, the time barrier space in the human body, where the characteristics reflecting the accumulation of personal evolution itself is regulated. It is recorded in the neurons of communication in every area that is affected by the actions of the physical body. The regulation of the emanations that interact with the energy of matter formed in beings from the beginning. The correlation emotional, spiritual is also according the capacity imagination, emotions communicate or interact with the cosmic vibrations. Cosmic vibrations mention because we are receivers of energy traveling through space toward our worlds and bodies. The vibrations are manifested in atoms, molecules and materials which give us qualities a variety of different reactions. We become aware of the forces that are unleashed within us. We create with these forces. We are breeding ground of the divine mind. Now and in the century, twenty first knowledge barriers are kept rising. Every day new steps are taken in the discoveries that give a more real than perceived idea and believed. There's More certainty than doubt internal processes, announced by the science. Soon this effort not be necessary, for minds that have evolved and left behind this time of fanatical obscurantism emerge as the new pillars of development for the future. I hope, humility, embrace the future of humanity. Young minds need know this legacy that is kept at bay by the systems created. In visual images, this gland that retain the emotions caused by the reaction to visual effects, effects of the spectra of white light that is phosphorylated in its processes, vibrates to the reaction of what is perceived, causing involving the range of emotions that encompass all manifestations attributed to humans. Laughter, fear, rejection, acceptance, love, admiration and all the attributes of human expressions, are an attribute of different arcs of internal harmonies. I say arches because it is a leap of sparks from one neuron to next, causing the effects of alignment and balancing interior creating waves that affect our endocrine system. There are even studies where it is attributed to this part of the brain, a region as old as the thalamus, fear or acquired the notion of protection against hazards. This range of reactions we experience are the interface between the cosmic mind and gradation that human beings can experience their relationship space-time in the compartment of the physical mind. It is a natural function that was developed from the beginning of time; It enables us to operate all the defenses and escape protection. The unknown sense, intuition we possess asleep within us, a sense that alerts us a spark all-encompassing and gives us an accurate notion and sometimes anticipated. As every human being who seeks unravel a concept or explain processes, you should go to the appropriate sources. That's why I go directly to the teachings that encompass the notion that advances the idea without having to produce. That's a routine that seek give something to know, because nobody owns all sources of knowledge. Summary of the findings of Dr. Esther Gómez Rosón published in 2005. Resting potential and action: 1. Situation rest: In this situation rest, the membrane of the neuron is polarized so that it is 90 millivolts more negative inside than outside. The resting potential of -90 mV is due to the concentration of sodium and potassium. Note: "Nanotechnology can cover deep into the system of energies that work in harmony with these membranes were formed to filter and decompose radiation creative thing field. only know a small universe of them, because it is a new and I am no scientist.

1- Mention logically should exist so I imagine countless discoveries, which are not accessed or on the contrary minds prepared to pay the scientific knowledge

2. Depolarization. sodium channels open and sodium enters the interior of the cell

3. Repolarization: sodium channels are closed, and potassium channels open and potassium leaves the cell to replenish negativity

4. Pump sodium / potassium: expels three s energy sodes for every two potassium. 5. Recovery resting potential brain hemispheres. Two hemispheres, one right and one left, which are separated by the interministerial slit. the cerebral hemispheres are covered by cerebral cortex and gray substance. * Note: "These materials were accumulated in the passenger compartment of the brain for eons of time, during the reaction to the energy that caused its accumulation and the need for material to neurons process that matter where basic materials were developed by attraction. compounds that provided the basis for the creation and distributed through the channels of the material body. Following the laws grouping of materials necessary

for the survival and channeling the energies that were creating gaps attraction. from the forces, necessary to complement the work which they were imposed to achieve harmony in creation. " The cerebral cortex has gyri and sulci that increase the cortical surface. Cerebral cortex

voluntary movements are developed. They become aware sensations. Information is stored. Mental functions are developed. Recental located in the ring, in the frontal lobe. They order the movement of voluntary muscles of the wide body contralateral originate. The fibers originating from the pyramidal tract. It is ahead of the motor area. This area program movement. Has many connections with striated nuclei and thalamus that act as hubs? Note: "Based on these observations demonstrates that there must be a hotbed of energy reactions, creating the environment for these processes are possible aggregation of this process have a maturity of millennia and strength that led to the maturation, should be so subtle that only the divine light of creation, can act in these processes. Its qualities are maintained as in the beginning or with slight variations. these must have suffered millennia of variations in material changes and modifications of conduct which were refined processes that possibly even science is trying to document. the energies that have created this scale conductive fibers for the variety of energies circulating throughout this skein of nerves and neurons that obvious complexity, does away with imagination. and this process is progressive overcome and create channels to modify internal processes. Being dual that energy goes into the process of the thalamus and branches for their channels created by of evolution area. Autocephalic is in the frontal lobe contralateral. Voluntary movements of the eyes and head are given. Area calculation, recognition of body schema, touch recognition Region and reading at the end of the Sylvian fissure area of language is almost always in the left hemisphere Divided into several areas: A Broca area Wernicke area frontally level -.... temporally and front of the occipital If the mechanism fails can cause: Aphasia failure mechanisms Language comprehension or expression dysarthria. error in motor mechanism speech organs Note: "These notions of alleged failure normal functions must somehow be affected by the course of a stream of energy. His interact with accumulated and was diverted by a malformed source diverted from its course and materials prevented the proper function that was intended "limbic system.

Region that controls emotions, motivations, emotional behavior is enveloping the body. callosum, the area of the frontal, parietal and temporal lobe * Note: "with the abandonment of the control force of will, many functions have been lost. The maturation

of resistance, abandoning overcome the will to face challenges. The correct direction of energy cooperation with these reactions, weakened structures overcoming barriers. The maturation of new networks diverted energy to overcome attacks uncontrolled pain anxiety and frustration. The lack of valuation sheds a form of neglect memory and attitude of abandoning the will to fight and recover those areas.

"Area of memory There is not a certain area. It has given much importance to hippocampus. Grouping nuclei .is next the third ventricle arrives sensitive information Participates in motor control maintains alert *

Note: "the memory of the brain does not have a specific field, as each cell the body is a brain power and has an own personality in their habitat and dominates a wide range of energies. Provides development and accumulation of information sharing with the swarm of neurons and nerves through communication networks that vary, as it has a fixed pattern of behavior and conform to the needs created by the same system "striated Nuclei.: formed by the caudate, pallidus putamen .and Globus they motor control Note: the colors of the substances give an idea of the power of energy that must absorb or filter, become a loophole vibrations have to regulate process.

Bring to a specific accumulation depends the specific role of the qualities that must supply like any other material that accompanied the evolution in each created thing. Hypothalamus Region consists of gray matter.is next the third ventricle, ahead of the thalamus. it is. the main growing center Note: Its function is to maintain a store of matter, be the supplier of the components necessary for the materials that will be then used in the system are accumulated to be conduits of creative energy are given. The brainstem consists of: 1. midbrain. 2. Bulge. 3. bulb. Substantia nigra midbrain, where dopamine necessary occurs in the control of movement. This area works in coordination with the basal ganglia. -third Cranial- core MOC par. Responsible centers of consciousness and waking rhythm. Core protuberance MOE and MOC Sextus cranial- par. It is the output of the facial and trigeminal nuclei. Bulb area where the pyramidal tract is crossed. Respiratory center. Departure of the hypoglossal and vagus nerves. Vestibular balance centers

Author. Esther Rosón Gómez Year 2005

sodium potassium pump ATP-adenine-triphosphate is a transmembrane protein that acts as a carrier exchange antiport -transfer simultaneous two different directions hydrolyzing ATP. It is an AT. P-type transport passes, i.e. undergoes reversible phosphorylation's during transport. It consists of two subunits, alpha and beta, which are integrated in the membrane tetramer. The alpha subunit consists of eight transmembrane segments and on it is the center of ATP binding, which is located on the cytosolic side of the membrane. Also, has two bilateral binding centers, extracellular centers and three intracellular binding, are accessible to the ions depending on whether the protein is phosphorylated. The beta subunit contains a single transmembrane helical region and not seem essential for the transport or for activity. Note: The energies thus described are part of the divine energy that began to create the varied matter being. This pump is a protein electrogenic as it pumps three positively charged ions to the outside of the cell and introduces two positive ions within the cell. This involves establishing a net energy through the membrane, which contributes to energy between the inside and the outside of the cell, since the outside of the cell is positively charged relative to inner. This direct electrogenic effect on the cell is minimal, since only contributes to 10 percent of the total electrical potential of the cell membrane. However, most of the rest of the potential derived indirectly from the action of sodium pump, potassium and is due mostly to resting potential for potassium.

Motion and rest, an equation of the notion of divinity. Note: Creating membranes of different thickness, due to the ability to filter and graduate energies to their proper role is in creating new substances and connections in the construction scheme of matter in the body. It is the most important role in the regulation of activities to supplement the body of a control emanations that penetrate inside. Precisely this factor allows the intervention of the creative will of imagination, enter control some of that energy for communication or often this energy capacity and enjoy a heightened sense of divine consciousness. Harmonize internal fluctuations of the matter that makes us refine its structure, conductive to open a channel between the conscious and the subconscious, the border of being with nonbeing or vice versa. It is essential that our being perceived that sense of inner harmony. V- emanations of the original United Intelligence, wisdom and knowledge's own will to create, wisdom that concentrates all things, intelligence to stop ideas by association, emanation and induction. Grace, love and mercy, be inherited attributes of the divine mind, the power of judgment always with us, perseverance until the final victory, greatness and majesty. The foundation of all the forces that have been activated since the first emanation of that will, finally, the kingdom where dwells the God of Creation and gathers all in its manifestations consciousness. Note: The Will projection, is the primary source from

which everything emanates, if it is the will of the creative force and its course was the creation, for that was at first that must have covered a moment in the divine mind represented in the cabala. Concept accepted by the mystics who diagrammed the divine order. To transfer a vocabulary not spoken keeping the keys of creation, that being able declare for the future without loss of content. The correct teachings, they inherited from the caves and ancient stones. The resurrection recreates the spiritual legacy of ancestors and was carried out in caves where he meditated for periods of time, avoiding the disturbance.

The reaction between understanding, intellect and wisdom, created the dimension of space and time (the greatness and strength of attraction) away from foreign elements, distracting the internal concentration. God created a vehicle to express their divine essence. The human being is the heir of that kingdom. Its creation was not spontaneous; It was through many stages of evolution until matured an expression that covered everything. The kingdom needed above the kingdom into being for levels of maturity and exchange of energies that interact and attract primary source for adaptation, the attributes to be manifest. For that we were created and function as the emanation of a higher consciousness, which needs a lower consciousness to realize it and take life in a plane that is not him and thus make their own existence, while it manifests and recreates itself. The beings did not appear moment running around the earth and made completely. The cycle of life, incarnation after incarnation, is repeated in all realms that is the existential law. By that time tunnel space is traveling the creative forces to comply with the laws of evolution and return the same tunnel forces

to regenerate the primary source. What was first manifested by laws of creation, the other follows are the evolution of that creation, a rise of the creative mind far as human understanding possible capture. It is mentioned and relates as a tunnel that acts as a communication channel between two sources that the energy travels per its capacity of attraction and repulsion, so that this energy not contaminated or attracted to others of lesser quality should be subject to laws of manifestation and regulated by their attributes aggregation and ripple. Reincarnation for centuries endowed the human sources of power and free action to mature new expressions later, returning to the original source, reintegrating the great divine mind. By cycles, these same energies back into a new body to continue growth to exceed its previous state or go through a degradation if the process is not exceeded. The power that created us brings us back to its parent to provide us with new energy and vomits us back to earth so that we prepare our path to purification. It is the law that was imposed, evolve. The demonstrations since the beginning of time, the physical creation, material nature and spiritual, with cosmic attributes of the soul, where the cosmic creator recreates itself in the manifestations of those minds that are harmonized with their spiritual laws. The purest emotions of being uplifting spiritual knowledge and transcends the material ego raises its essence the cosmos, or the call God our consciences. Uplifting and mature that energy and reverse the God of creation enables us to be bearers of spiritual light.

Thousands of thinkers we admire in schools for their wisdom, the importance of knowledge takes us to the innermost recesses of our own nature. What comes from the divine mind through the human mind corresponds and has processed through this center of creation, home internal processes through being. Everything created from all branches of knowledge, what we see in the history of mankind has manifested under the same universal laws. Just being aware and knowledgeable processed everything that exists and created by man. He has also taken of what surrounds the cosmos in the varied motivations of nature. As I said before, we are at the gates of knowledge that had long ago surprised the common man. The last triangle of creation, glory victory, they create the foundation for the generation of new beings and their powers of procreation. At some stage, what was a perfect triune manifestation it was divided so that the forces be harmonized again, when all were one. To that end the reign of divinity manifests. There is hope in the preaching of Jesus being by evolution must find the way to return to the divine nature that it, created what he called, eternal life. It is a challenge and it follows that creation took place on this earthly plane. By deduction this is the plane of containment where conditions for holiness, or the holy spirit that proclaimed Jesus Gathered. If the divine plan to create at this level for some need for expression in this plane came to develop the kingdoms that currently contemplate and what can be captured in the future. Jesus is exposed by updating stages of knowledge of the great initiatory school's class, carried a simple projection for minds to which it was addressed. We must conclude that is a plan of creation itself. We are the winners of realize that we have the unique notion of divinity within us and we must look devoutly that opportunity. Human heredity and evolution Human beings not descended from monkeys or primates, as discussed trying for years. They are great theories that have been developed by great minds of humanity. Like other knowledge has filled a void in human experiences. The projections in human nature in its material basis is the only way that has focused the functioning of the human being. Science has been commissioned to rectify those errors of assessment, which, although not far from many realities do not fill the empty complex. If we look in the annals of antiquity, there an understanding, even if metaphysical or esoteric, of creation; It was at the time a product of human intelligence that was closer to the events: creation. In coming times, more light on these issues, and possible the last link is understood be thrown.is my goal focusses the creation of beings as something akin to the superior forces of subtle communication skills graduating from being with its creator and is related to the attributes of what generated. The designation of names or concepts is a way to give a space where the concepts come together for a universal understanding of what motivates human to be overcome and continue exist. If someday that inner legacy that is common to all is would be the end of an existential harmony distorted. In the kingdoms that do not contain our notion of something higher thereon differences. Other creatures were developed on par with humans. The human being developed, he was involved in the same struggle for survival, to overcome the ferocity of the other, with superior with them. The fact that their skeletons are mixed in their own environments, and were in similar structure is not proof that one was the heir the other; simply they shared the same environment and the struggle to impose on others. The fact that its characteristics are like human being does not grade line of other beings to equate human being. Reincarnations give the man of that special feature, because returning to body are the divine attributes of his soul personality that would not be present in any animal. What has been called soul is a specific function within the tangle of emotions that mature within our being. The soul is the quality within the subtle energies that are stored and encompass the whole, in our inner emotional being. One area that includes the vibratory rate and stores with all the emotional intangible attributes that evolve and mature for the period of existence on the earth plane. A quality we handle and project as a function of gratitude to God of our consciences, while separate us from another species. We are aware of what we call God and all the attributes of divinity operate in greater or lesser degree within our being. The difference is that we realize and mature willingly. All this combined with the variety of physical characteristics that make an internal world difference and physically. It separates us world of demonstrations that we must understand before declaring that descended from other animals, whatever they may be. We are a manifestation that follows patterns and different laws. Studies

show that communing with features of creation in the beginning, aggressiveness, idealization, concepts, innate intelligence that makes us superior, the realization, imagination, consciousness of things. For centuries, the knowledge that accumulates in the memory of the evolution that has cataloged files (Akashic) by mystics to all world standards, human beings from other creatures that exist apart. Best My knowledge, this is a source in the cosmos where are recorded all the knowledge that mature soul-personality and consist of knowledge, emotions, personality data. The number of atoms that formed our physical being, the exact knowledge of each attribute of manifestation in each volume or cell of our physical and spiritual body. As The human being it is concerned and its counterpart in the source where maintains an archive of all creations that have emanated from his own being. Furthermore, the DNA strands (Genetic components of each cell) in humans would have to be the same as those of animals, but in the same direction DNA testing varies by 2% of the primates, which aims link our evolution. I must confess that I am no scientist nor possess knowledge of genetics. If I make a mistake, I hope arises a prepared and pay these ideas to establish a clearer understanding mind. You may already be cataloged. If not so, it should arouse interest for proper scientific research. I imagine that book in question must be a lot of authors and theories based on this same subject. I do not want confuse the reader with theories that do not contribute to something simple to understand. My perception of these which were far from what common being to internalize. Meditating on that I have tried to simplify the issues addressed, although it is difficult communicate something that is so easy, unravel or rescue stage where he has been buried by concepts imposed for thousands of years. Calloused minds by the same concepts, has a challenge of organizing and accept new arguments, this is nothing new.

53-Creation process in saltwater

My express knowledge in this book, is a story of mystical type, alchemical, metaphysical, because I personally attach bring this knowledge to the greatest minds, from ancient mystery schools, philosophers, doctors and mystics, including the most mystical greatest scientist that ever existed in his time; Jesus. The teacher tried to direct the course of their people to freedom, physical and spiritual in his time. In other stages of evolution, there have been bright and mystical minds that have endowed humanity above teachings, for the advancement of the least prepared. My way of expressing this account of my experiences through daily experience is looking the relationship between the true path to truth that dwells within us, the same as reincarnation after plays reincarnation as a portrait of the soul. This method of gnostic schools where every student is prepared to travel to the ends of their inner being, descending into their own hells and mature spirituality to understand God and human suffering when I degenerated being. Crown Paradise Escondido

Being through its evolution

The energies of the combination of sodium, potassium and other chemical components that come into the functions of creation, interacting in the center of the thalamus, in our central nervous system, through the endocrine system, thalamus, pineal, hypothalamus, hippocampus, secondary glands communication and emotional as thyroid, neurons and dendrites, are rooted in the principle of creation. They have their own DNA strand that has changed since its development in the beginning of creation; simply they have been transformed and expanded as a mirror or a source layer or membrane. Produce energy where primary cells derived their structures, because they depend on these substances for their survival, because they are the source to produce proteins and their derivatives and aggregates. Without them, the first cells could not survive and structure of which led to a concrete being originated. Like the planets and bodies in their orbits they have a belt or mantle of energies that interact with matter that form. Thus, the human body and all its sensitive parts are surrounded by the mantle or aura of energies corresponding to the qualities of the materials that are formed. Very important define whether that is an external irradiation, bodies or an approximation of attraction is present to second reactions that are about happen. The laws of attraction have that quality as in other realms, use the elements that attract complementary. This field is the immediate aura of the thing created and can be seen by many means which equates to the seven colors of the refractions of light, pure white, original creation, the projection being to not being, the principle of all. It recreates an individual magnetism that is the sum of its composition with the spectrum where stated, the total of the matter is what defines. The characteristics of the first cells in salt water, have their environment as result that all forms that were developed initially had a common environment. At the same time, their organs were developed, evolving into what other creatures were emerging from the primary form. As result of that common principle which everything emerges, many of these creatures spawned or expel their seminal fluids and waste in the same environment, and being inherited from each other, ova or eggs could interact in their primary development functions, attracting or rejecting qualities. At some point, they must have arisen creatures of all shapes and sizes, which interact together, by genetic selection survived the most capable. Currently, the behavior of sperm to fertilize the egg resembles the conditions that had to overcome in the water in that ancient epoch to move and reach your goal. The same conditions are present in the play within the self and many of the realms of creation. The human body is a replica of that process that occurred in the salt water and was encapsulated in what would be the evolution of the species, which gave that quality human being. The endocrine system began to develop

by the need for primary material that was common in salt water and which was now in contact before emigrating to land. They were forced to migrate in search of the necessary components to manufacture proteins and substances necessary for survival, so began the migration of salt water to earth and earth to salt water over a period of adaptation. Similarly, external parts were adapted to the new challenges where limbs and other organs that were modified to meet the requirements of adaptation arise. Emerging from the seas and salt lakes, the system began processing the thalamus sodium and potassium directly from the emanations of the Sun, another key to life. In that process, I think what was a skeleton to support graduate sensitive areas and their ability to emigration. That created the diversity of species and the formation of being. In the beginning, primitive forms, and create substances such as proteins and other primary aggregates, she suckled bottom sediments that accumulated by the actions of evolution and climate phenomena. This provided them with materials of different compositions to develop new areas of expression. His interact with the same primitive forms, which began to develop materials that are added and depending the environment where they are located. Not all forms that emerged had access to this area. The brain that developed acquired unique conditions and regrouped around specific subjects. Today they are channeling their heritage of antiquity. If this change, evolution would take another turn. For this to be understood by less prepared is logical that the disease is given by any imbalance of the natural conditions of how it was created.

Doctors who provide the cure of disease, simply seeking some form of knowledge for restoring material to be a substance which longer produced in the body, or decreased its level of stability. Everyone being is its own world and its attributes are the sum of its history of evolution and the elements that have been accessed. Both in their food which adds new material for the formation of what will be its structure. Note: -. In my research found the information on the findings of a scientific twice Nobel Prize in biosciences water

Doctor Linus Pauling

In his studies of biology as water science has discovered that every 1 (liter) water sea is composed of a marine soup containing; 965 cc water, nucleic acid more DNA, essential amino acids, proteins, fats, vitamins, minerals; (a total of 118 teams from the full periodic table in addition to phytoplankton, zooplankton-krill -. Omega3 eggs fish larvae, carbon chains, matter particulate-. All this related to the origins of cellular life and to my surprise, states that seawater is complete nutrient nature mentions that adding 3 parts of freshwater to sea like human plasma solution is obtained.

'This shows that the human body has the same elements as the original salt water, with a slight variation I observe of comparison. it means that we seawater a copy of our material body. All components are related to our original physical structure. in fact, it conceivable that that arise in the beginning and current adaptations, the body has changed some of its characteristics, its environment changes. we're not exactly what we were in the original encounter. Some features have varied in composition and use. The variation of the human plasma used for the test is the current that was combined with fresh water to recreate the variation of the emigration of being salt water to the land where we eat fresh and processed salt. This added to other tests check my statements that humans originated in salt water. Gnostic knowledge is the only internal source to harmonize our rational being with mystical and divine attributes, who live by spiritual heritage within us, the maturation of our soul and personal conscience. The intuition of God of our consciousness elevates us to a perception of states of exaltation upon contact with that divine harmony. To emigrate to land faced a period of adaptation and interaction with new direct emanations and subtlety of vibration to which it would have to adapt. The phenomena that were directly exposed. The causes of the variation begin to interact and the needs they would have to overcome was very different. Thus, the new expressions that them to survive need cause rise. As for the shared life in saltwater, countless species developed a complex system by the need to survive. They moved from sex to continue playing their species and not make this go away. Cows are heirs fish, Anthia, bivalves and mollusks. I read somewhere that the adult hyena also changes sex, there different methods of reproduction. And of course, our thalamus gland lived the same environment and evolves superior to other species characteristics but

similar in shape created, which provided with the sixth cluster of attributes on species: a dual, trill, and androgynous egg both. A comparison of the evolution where being was present. In humans, the sexes are manifested one at a time, the other sex is present inform asleep. Possibly, in another embodiment the other can dominate the stage of creation. As I explained earlier female hormones determine the sexes when mature, the area where the clitoris is manifested in females is the same area where the penis is manifested in males. The attributes are contained in the speech area of the brain, and for me the thalamus, which is the gland that defined and recreation for first time. There is no way to declare or prove that the human being was created with like other beings that originated at first features. There Many stories in different ancient civilizations, which focus the principle of human beings as an androgynous egg. Your body may share an evolution equal to the other and then divided into male and female. Even today these phenomena that appear are numerous rare in nature, something natural in the continuing evolution. On the human side of androgyny cases have even been scientifically cataloged and are being studied are repeated, but within the same body. In one version of Reader's Digest selections (Page 194- The wonderful human body) - ISBN-84-88746-237-- I found this statement that does away with this maze of imagination. Female hormones are responsible for the maturation of organs, sex and manifest themselves in the clitoris. In the female clitoris embryo is the part where the male penis develops. In development is the same where a joint body is developed to express the two conditions in the human body and manifest themselves in different parts of the embodiment of a sex that determines which defines the expression of DNA in the future. In areas that clitoral arousal in women and foreskin in men is due. This generates a scientific explanation of how the sexes do not need a different body to manifest the dominant sex. The laws of the God of creation made things perfect. Being must marvel in the realms created because in them disclosed the divinity of creation, and I'm- I personally thank this sublime being, enlightenment given me on this day. In the human kingdom beings that developed in different areas of the planet they inherited his physical way to survive conditions of nature itself, which had to face for their survival characteristics. The variety in feeding each other, made the difference in their physical characteristics. While one of the races was devoted to hunting, there were others who were not under this same influence or separated by the rise in different areas. By mentioning this detail, I assume that the purity of their food capacity with the purest qualities of life and its development was more advanced in terms of their physical form and inner development. Another detail is that when storms arise and the different phenomena of nature, the original forms were dispersed in the water all the planet due to the same phenomena. I Look at our American continent many of the plants and animals and being's other species have been produced by these natural phenomena. His brain was adapted to the evolution of materials, such as fruits and herbs, pure water that is possibly consumed fishing and thus the development of cell characteristics. One very different from those that only fed on hunting animals and their DNA was mixed internally with the animals, acquiring physical characteristics of the meat they ate, like his behavior to be linked to beasts and animals that had to dominate by force, their attitudes were of this nature. The absence of a rationality that made them overcome the fierceness of the animals endowed with such attitudes and mental and physical evolution. Other hand, there were scattered races that adopted different development phenomena and were more interested in nature, behavior and the ways in which this manifested itself. Their emotional attitudes were mature differently and their behavior was more friendly among his people. In other words, his behavior was more emotional. Somehow, they adopted more organized forms of life and respect for nature itself. Natural phenomena were attracted and sought an explanation for everything that happened around him. At some point in the evolution nomadic tribes of Europe they joined for some reason either fight or subjugation or dependence each other, because somehow reasoned that joining forces was easier survive them. Some were diehard hunters, others much were the victims of others or the same wild animals. Another reason must be the phenomena that occurred prior to which they were exposed, more natural phenomena did not end with all creatures that evolved on earth. Apparently, there was cataclysms, created by cosmic phenomena or phenomenon lock somehow emissions energy from outer space, for a period and the creatures who did not achieve development in their endocrine systems, which enables to manage that crisis, they succumbed. As cataclysm, I mean the blocking of the emanations of the sun, interference planetary phenomenon, which opened the ozone layer in any region of the world or immersion of the earth in a gravitational field of an asteroid created a change in the Earth Gravitation or rare material affection with our environment. The causes of variation in behavior or development of people. The other realms of creation are influenced by the energies we process the characteristics that led to the emergence the phenomenon of life. Should by reasoning that any external phenomenon that change, of any kind, alter the laws to which obeys our development, whether physical and emotional, obeys the same laws. Imagine that only isolated physical phenomena can end the whole life would be a miscalculation. The violence attributed to cosmic laws only this behavior, verging on little imagination of our creation and evolution as if it were physically analogous to existence.

Beings created by the primary emanation obey universal laws, divine as they are by the emanation of intelligence that carry within us water.

Effects of ionization in the salt water

The mystery of creation that only occurs on planet earth to the present. The cabal of creation: For my little knowledge of Kabbalistic principles to which I had access shows that the notions of kabbalah apply directly to the creation of human beings. It is a way to give meaning to what is known and disclosed since he is aware of what is experienced and perceived. Are the statements attributed as a god or supernatural force, influence on what was created as a being? Is our internal process of creating a single intelligence in the universe? It is possible that races from other planets are interested in our planet, it would be for the sole notion of how intelligent life arose in him. We Faced with this question that the principles set out in the labyrinths of kabbalah should work in all creation in the universe, not excluding extraterrestrial life. A condition that follows the cabalistic statements is that the forces emanating from the creation are specific and obey cosmic causes so far to accommodate human life as known so far. If emerge creatures with a different way of interacting with the universal matter they would have to design a cabalistic system that is expressed on all laws that are manifested in the universe over all the kingdoms of creation, thus harmonizing knowledge with the existential reality. If the infinite universe exists in other areas of underworlds or areas where galaxies are an alien prototype to our life, where other demonstration system develops. That physical existence not necessary, only be a flow of infinite attributes or essence of manifestation. I mean the realm of with cosmic science encompassing, where the refraction of primary matter arises. Effects of ionization of sodium and potassium. The functions of the gland: thalamus is possible that intelligent life on Earth is unique in the universe and if some intelligence from outside our galaxy will approach our planet would be looking for the frequencies that are combined from outer space and they accumulate on our planet. Only condition: Being a single distance from the sun and the combination of elements emanating from the different sources of energy into what up the planet earth that has been added since coalesced the raw material is a unique condition and we possess. Our imagination is raised to be prepared for actions of races from other worlds, is logical think that may be superior intellects, their lives be very different in molecular structure to ours. We are a unique condition that derives from the energies of sodium and potassium and elements like oxygen, carbon, and elements in the salt water, which is nothing than the concentration of the sun itself in the water of the oceans we They surround. In addition to all the elements that have been concentrated since the emergence of the universe in the material spheres of the earth itself and its ethereal fluid such as air, the energies that encircle us, while we are exposed to the energies of other bodies credited with its unique conditions, the way of life and phenomena that we are manifested, produced by external forces. The function of the first membrane (Thalamus) of creation that gave rise to humans is one of the main goals of anyone who wants copy our source of life, the origin of our race. If someone wanted double the life in outer space to originate creation processes should follow the examples of divine creation. How it began? __ What active ingredient that these elements are organized and start these reactions? __ What energy materialized and brought the qualities that gave start to the first cell? __ I analyze this in relation to human. establish that the creative energy interacts in unison with the different forms of life emerged. It was a sprawl of some source that emerged suddenly and generated these reactions, initially, by disorganization. If not who would give an idea of how it happened, if it was not the intelligence of God in being created. It must have arisen in increasing degrees. His qualities must be vibrating in its concentration in different areas of the planet, it led to life were arising in a related one with the other principle. Begin you interact with various materials, depending the area where he said he reacted this energy to the attributes of the materials were added, creating diversity. This was determined by energy frequencies that attracted and combined in their structures. The different temperatures cooperated with the generation of forms. Speaking of temperatures found another link where the very land from himself, as a mess of nature, contributed its primary gases and materials to help the delivery of energy traveling to impregnate. In block features that ultimately emerges as a human being, they reacted molding materials of different characteristics of compounds or atomic variation. Electronegativity is a sensible explanation for the reactions that occur in the field, itself, is the key

to our spiritual advancement. In fact, that subtle energy that excites us and enables us to realize energy of divinity, beauty created in all realms, is direct communion with the light of God's Kingdom who created us, the power subverts those energies on feelings or real images. While having all the characteristics and qualities of definition in each assessment, scientific minds seek accommodate the exact words without leaving an opening for speculation. The saga is that if you test something, we realize that someone patented later that same observation about advanced data science. The salt water was already concentrated solar rays that should be weaker and not penetrate fully the layers of dense water, it was by freezing or turbidity. In addition, the atmosphere was denser and hinders the penetration of light, which grew progressively within a framework of reactions the solar system.is possible that the attraction of atoms and their internal spaces create the perfect material for this approach, the sodium and potassium content in salt water serve as the primary elements. From that moment of creation that phenomenon has not ceased. __ What emanation creates this first manifestation on Earth? __ What evidence continues interact so that the attraction continues as a single act of creation? __ What qualities for millennia have been added to the original attributes that awakened life? Neutrinos, __ What role have on the spaces and spectra of matter? A blanket of subtle energies with its known space, where everything travels to all corners of the universe. Wherever required its complement there is mixed and interacts with the related matter of his qualities. Specific mind functions, imagination, intuition, files memory, emotions, frequencies where they accumulate and can be attracted to emptiness of mind that is activated to collect the previous legacy, needed to complete a function or retrieve data. Specific The mind of being harmony with the natural function of some minds that meet that information to their advancement and profit. What particle or combination of particles against reflected the other as a mirror light refraction, and will equip the darkness of light patent invested in the speed of light and darkness creates. What kind of darkness? An empty manifestation of mental clarity, images supplanting time information and a light beam crosses thought, an encyclopedia of information resurfaces and mind can spend hours retrieving information on a specific topic. With the focus on detail specific that will join a chain of related information. But it is not the manifestation itself, is the chain of events that crowd and attract other creative powers that bind the rest of the elements associated by some order we call divine or harmonic the first stage of life it manifests it would then humans. The same evolutionary conditions in what would the variety of the kingdoms that arose and its derivatives, to overcome the pitfalls of physical reactions were recreated. The first manifested as a primary law is to protect this source of energy that begins to act, with the things that surround and with which it interacts. Arises an outer layer that protects from damage becomes dense or lighter, depending on the radiation itself graduates and controls for interacting with the vibrational rates necessary to generate their structures and functions or vice versa. It is the reaction to the powers that penetrate to promote their influence in the matter: the generation of the original kingdom was implanted itself in the thalamus God as the divine substance therein earth. The insulation of the creative force emanating from the beginning was trapped inside the cabin that was closed to the influence of external forces. Only those that were necessary for their manufacture of its components. "The Bridal Chamber" -The Eve and Adam were prisoners of outside influences. In principle paradise, then degradation kingdom was embroiled in pure being distorted himself and strangled, his body was divided into male female. The loss of contact with that pure energy and a unique condition that the ancients discovered, loss of alignment with the forces of the same system that house. The alpha and omega: Creating Start communication across networks emerging from its center, nerve communication terminals. Arising from the need to create their own environment and its components. That first egg creation binds of proteins that cells begin to produce; process takes some inner magnet attraction or what we call consciousness or individual intelligence, or need attraction to other elements that provide electrons and complement their development. That quality or notion of matter and its components, is both divine intelligence that emanated toward creating and isolated herself in a cabin where he started control and stages of creating your own being. Countless items that come to organize this process in stages and adding time, as needs arise to create. An energy that envelops in magnetic and gravitational fields gives this primary creation, comes the ability to move and start moving in their environment and develop increasingly complex qualities. As the seed, each time you fill a new addition copies of some form to its composition and keeps it as the chain is formed in its evolution. So, it creates a copy of energy and duplicated in any other division thinks of herself. Copies shall inherit these genetic codes to divide, what is the law that formed, pass to their duplicates. That way of splitting an interface that defines a realm with unique features is created. By deduction doubles every created thing itself. Emerge at another stage of manifestation with the above characteristics or a clone of its predecessor. If that is the event of evolution because reincarnation is denied, and who denies. They need move them create bulges outwards. At the same time, the gland itself provides them with sensory branches of themselves to detect nutrients and forms that must evade and overcome in their movements. These features add an ability to copy and retain a memory of forms and energies of the detected in the future for catalog and identify experience. The same that would be recognized if manifested again and the reaction to the recorded information and the notion of choice or rejection of each to maintain a code to play as necessary to overcome a condition of growth.

At the same time, they should face substances that are added to food body to adapt their conditions to train the system to produce more cells with those matters that are of different qualities. You may eventually be much the variety they must modify their behavior to regions and customs of that feed. By becoming the processes that must be overcome and additions that come them are very weak by the concentration in the water, as all beings more complex created start changing their senses motion and create what would be the eyes to sense light and movement. A skeleton that supports the ravages of weather calamities and wave to it answers. This would use the same scheme shells to protect sensitive organs. Just as the skin protects sensitive organs and parts, while acts, filter for substances in activity. Internal to support their needs more creative energy that endow an imminent need to generate other waves to provide adequate vibrations to create the components that enable to develop new areas of mutation and development glands. Suckled fund nutrients were adding to their structures, in the same way that filtered fluid to retain the components in water, would benefit for creating both totaling a new feature to the above. They stretched themselves on their nerve fibers and became aware of the dimensions encompassed by the signals associated to new experiences. They were creating new nerve areas and cellular structures of the materials sucked to store that information and hence the first brain emerges. In the cells, themselves duplicating the genetic information, they passed copies, which survived did the same. Already a cell with two divisions and functions, each side control has its own genetic code. Internal sexes arise where the needs of transferring genes from one cell to another for the division cycle is completed, with the right polarities are developed. Internal primitive gonads arise within each duplicate to train their codes of evolution. In this process the invasion of other life forms such as bacteria or viruses and foreign to their ways by which development should start developing defenses for protection arises. This interaction begins the internal creation of what would the prototype of the human being and an intelligence or accumulation of precise details of creation. Abuzz with nervous reactions develops a central neuron, which responds to the impulses that receives and sends energy responses that are related and recreate a condition to correctly interpret the impulses it collects. By the need to accommodate a structure that was developing its size was doubling its structure in what would be the body and through its neural networks he sent the necessary orders construction of the attributes that achieve reproduce its primary structure in a body. He gave communication networks, where it reached full control of each cell created inside. The instructions sent through the neurons or nerve centers created to maintain complete control of every action that given in your body. He maintained and maintains a notion, as an exact copy of its contents. If any cell or group ceased its functions or destroyed, matter to replace was immediately generated and sent to right place. He created a center or cell factory in what is now called liver, which in harmony with other related subjects fluids creates the exact component to service the needs of active creation. Somewhere in his brain, he developed a unique ability to provide pulsed orders to make duplicates of cells already created. As always, in their primary cell duplicates or copied a replica of their progress in dating his new creation. It arises a palpable fact that this group of nerve cells, no matter the region that has been sent, forms a network of internal communications. Cells are aware of the functions of the other, forming a universal consciousness or mind what they are and their relationship to the whole. That intelligence is part of the control center, the thalamus and endocrine glands, which accompanying its functions. It is a perfect complex operation. How can a pain impulse take an accurate statement to this center and give aware of any injury or illness, in which cell or cells there is injury or condition, exactly the material required for prompt repair is dispatched his laboratory to precise site where is needed on the precise amount, directs all its channels to the places where the damage is, and what exactly double the previous? But there more: if for some reason this material, either by a ruptured vein or capillary leading it, the stream of liquid that directs you to your destination appears in a part which is not recognized and does not reach its destination, immediately the brain sends new to create a layer around cells and isolates, creating what we call cysts. This central seizes the knowledge of creation and we are aware, we realize that intelligence; alerts us and keeps us aware of their duties. That same area, thalamus, is the part where the profiles of all that is perceived is recorded and creates a map, such as a GPS worlds and experiences that have accessed a memory bank everything that he has been exposed. It gives us the freedom of movement and to choose situations that are harmonious with their knowledge. If Encounter any difficulty, we alert to avoid or generates the necessary impulses to adjust and overcome. This is an imaginary sketch of how one of the kingdoms emerged, the most complex creation: the human and the divine light, that produced. Memories that adds intelligence to everything you do and somewhere is recorded as an inner world that takes knowledge or awareness of his being, he realizes everything and can manifest. The concepts and stereotypes of human knowledge, whether in the branch manifest, cannot cover the entire plane of creation. Behind all this notion there are worlds of knowledge that are not and can be grasped by the human mind. We can lock in laboratories and see each individual cell and break one by one the parts, either atoms or all its particles, either in energy or its derivations, the way that each person wants study the composition of the bodies and understand their aggregates, their DNA strands. Of one thing, I am sure: that none could give a description and a concept that covers everything involving internal and divine of being processes. Gnostic knowledge is the only internal source to harmonize our rational being with mystical and divine attributes that dwell on spiritual heritage within us, the maturation of our soul and

personal conscience. The intuition of God of our consciousness elevates us to a perception of states of exaltation upon contact with that divine harmony. The wisdom that was present in the first movements Essenes and the mentality of that period could not sustain poverty preparation and imagination, where the attributed deposit was soul behind the ear and things like that. The biggest paradox, if you can call, for me to prove that after this first cell come from the combination of sodium and potassium and other aggregates, then, by evolution, resulting in the thalamus. Therefore, if being comes from this first creation and dependent sodium and potassium elements concentrated in salt water: __ how is that leaves its first source and continues later around the Earth, and then depends the fresh water to continue living? As is known, when setting sun arises fumes were possibly highly concentrated and elements accumulated in lakes and seas. The water, filtered by lots of land, is purified, while being developed and gradually was interacting and migrate gradually between salt water and land, fresh water and then serves subsistence. Not only humans developed this quality, other beings reached the same stage of evolution as human beings. The logical explanation and was not to the human mind was that the emanations of sodium and potassium necessary for survival came from the Sun itself. So, that there a correlation between sodium and potassium that are contained in the water and sodium and potassium emanating from the sun, like a cluster of elements that focused on the different atmospheres and subjects of what was the land. Some transformation arises in the membrane to accommodate both environments and adapted to direct ambient oxygen out of the water, the membrane that holds this work during the period was submerged in water, gradually disappears when it is adapting to the life outside the salt water. This results in material life that way possible both on land and at sea. The detail that strikes me is that in the beginning a flash of creation where the first outpouring of creative energy, a phenomenon that has not been repeated in the face of the Earth was born emerged. Everything that came in that short period simply said. That manifestation all forms and functions inherited creation in the kingdoms that preceded emerged. The kingdoms created just what they have done is a first demonstration evolve and varying stages of development, overcoming the first stage of creation for millennia. Simply, this creative impulse took office, with the attributes described every part of it generated, the qualities that gave each created thing from the beginning. The original purity of creation has not been repeated since it emerged creation in its beginning ... This deduction comes the argument that, if the same condition emanates, constantly and the source, the elements that came together to what exists, would every day creating new life forms. What we observe is that the original forms evolve more advanced states of their own original creations. Creatures and plants emerge from their predecessors, inherited conditions that have already been part of a previous life. These new developments almost double previous preliminary characteristics. But how come of nowhere something totally new features, showing an unknown being? It has not been produced in this earthly plane, as known in the creation and found the DNA strands that follow a hereditary pattern. What caused these chains of inheritance? __ What substance or essence the first creative impulse was generated, the thing, as you call, I interact in the moment that life arose? __

Chain emanations What holds us together in a common understanding of all phenomena clothing everything is known? Where the gradation of the most diverse emotions that characterize us as human race? __ What scale can classify and measure these vibrations do not obey material laws and projecting into the sublime scales of what we perceive? Why the infinite variation of these scales may not be fully grasped by the human mind? __ What is the harmony of creation and what belongs or is equipped planes, and why and what for? We know and are aware that directly affects us, and we not imagine their true functions. Our imagination creates and is part of them, but not dominate. We attract energies of cosmic plane, come into play with our energies, which we call spiritual and physical, and build within us, we not aware of their specific functions. Just we know work, but we not classify these functions to capacity. We use phrases, verbs, all lines and even develop a clear idea of those functions. If Leave, they continue their duties without us noticing. An interaction of energies at all levels, combining the products processed by our glandular system and the product is added to our physical body at the time part of the frame of the cell and energy structure of what is an individual personality. Such an exceptional set of energies of all kinds that not and are present in simple or complex life, tarnishes us understanding. Simple question: we are at the gates of the most extraordinary revelations that just capture in this paper. In the ancient schools of learning countries, India, Egypt, Greece and Mesopotamia teaching methods endowed disciples and teachers of higher knowledge in the realization of this knowledge is used. How man, through voluntary processes, could influence the processes of creation and retrain your body and mind for higher functions in managing these energies? Laws resurrection, astral travel, the spirit flow out of the body and Return- reincarnation, where perceived that the new being inherited from his former life personality soul qualities that not destroyed or lost with the change. It showed that a new being manifested something of a former self and simply followed its evolution to more advanced stages of knowledge and maturity. It became known as the esoteric Hermetic schools or mysteries.

72-Because diversity

After millions of years of living and have developed a complex advanced system and where vision systems arise from the emanations of light and many qualities they were perfected per their survival needs, the fact of living in an area of water he trained in ways to breathe in water for increasingly shorter periods. Interacting water to land and vice versa created

other alternatives to survive. In this being manifested in the placenta, where the creatures are developed. This reproduction system has never changed until today: being is kept in a bag of fluid. A membrane designed to protect a body to the same capacity as the first cell protection which is contained creating a duplicate created. These are universal laws of the created and the conservation of intelligences, that he created, has endowed being of such delicate functions as its own emanation. Process like of an earthworm food. At the time, the earthworm has more than one -some, seven-heart and is a simple model of what our internal system to process them. Other nearby systems complement subsistence, such as the lungs, which becoming more able to filter the air. In the head area, a membrane is manifested in the newborn child who breathes through it for a period of adaptation. Its existence model submerged in water. They have done experiments where the creature can remain submerged without breathing through the nose or mouth after birth. I even remember my mother blowing this area child for choking. This gives us a notion and proof that we still have vestiges of the changes operated to migrate from water to land. It is a quality that is also evident in another known species.

The path of the Divine consciousness nervous Each strand runs through the confines of our physical being, establish communication networks most perfect you can think the human being. Penetrate each terminal and every corner where some function is in process. They function as the roots of plants in nature. Its function is occupying a specific area to establish another center of sensitivity, where collect information and return to the command center, which is the brain, the thalamus and secondary glands. But this does not end there: they are responsible for carrying data and instructions for activating the sensitivity that gives us a protective reaction, in case of danger to all components of creation; in addition, if some injury or illness disturbance activate other control centers to generate the necessary to establish harmony and preserve the laws running at full capacity reactions. Without such control, human beings would not have survived the amount of time he has achieved with evolution. Only our subconscious is aware of the activities that can access internally.

A clear example of the power of these internal energies that can manifest is related to epileptic seizures. Discharges are so powerful that make getting nerves and muscles of the body where the affected contortions manifest. The same reaction is observed in meat receiving an electric shock. The entire structure of the brain derived from endocrine glands that respond to the needs of survival, plus all the complex functions that has developed, human beings, controlled by this small communication center in partnership with the thalamus and energies that penetrate our system, charging of atoms and reactions that do not control, traveling to every corner of our body and out of our body. The first cell starts to agglutinate a mass of cells that form a mass or body is then converted in the thalamus. Develops a kind of foot or opening sprouting from a central mass that used to excavate and secured. This protuberance has a hollow tube he used to navigate and detect movements of other species, simultaneously, to suck matter, combining with proteins to keep creating what would be a body. They developed brain, pédicas and visceral mass. Through the hollow protuberances nervous system consisting of neural systems and networks developed. interacting He then created an extension that developed in what was then the body, an aggregate mass to form a more complex structure that would supply their needs, where he created a structure to respond to their needs and in process of creation, being. The heart, in a transparent bag, has pericardium and irregularly atria and ventricles. He developed aortas, which carries blood to entire body; the blood returns to heart through poisonous system. This small scheme of evolution, acting seems the development of human beings far as nature could express, in principle. Our evolution must have been a different and more complex process, but the processes of evolution are consonant to examples like these.

Claude Allegre

The findings of this French scientist who support the theory. Experiment "Black stripes sun." When darkness comes light or (page 73: A little science for everyone) are complemented Newton had shown that light was the result of the combination of the seven basic colors. Refractions through prisms break the white light that enters our system in a spectrum

that reflects the seven colors of the rainbow. In the experiments of Bunsen and Kirchhoff to show the refraction of sunlight and why the fragments of light to dark occur in the solar spectrum, to throw dust sodium to a flame and analyzed by a prism emitted color it is verified that sodium is equipped with characteristic lines, well defined. By doing the same experiment with potassium, they found that the stripes exist, but they are different. They studied all possible chemical bodies to analyze the chemical components of the Sun and, later, of distant stars. Thus, born astrophysics and cosmic chemistry. In the experiment of Fraunhofer identifies the black stripes of the solar spectrum, what was its origin. Fifty years later, Bunsen and Kirchhoff noted that among the rays of the solar spectrum, two of which correspond to lines of sodium that had been identified in the laboratory. So, they decided out front of the prism that sunlight analyzed a flame with sodium. To his surprise, the two black stripes became blacker. Kirchhoff then advanced to the hypothesis that the flame sodium had absorbed the solar rays emitted by sodium. They found that sodium emits and absorbs the same stripes. From there he went to another deduction: if those same sodium emission lines exist in the normal solar spectrum exists between the prism sodium absorbs those same lines. Attributed sodium absorption into Earth Atmosphere. It would be a mistake, because if absorption stripes come from an atmosphere, that's the same sun. In other words, if the solar spectrum contains black stripes is because the chemical composition of the solar atmosphere filters, removes and absorbs certain rays light emitted inside the sun. the sun emits light and, in turn, its atmosphere

absorb part of it. Therefore, solar emanations to land and water include sodium and potassium are concentrated in the two bodies.

75- 'heroic age of spectroscopy

then reproduces the chapter "The heroic age of spectroscopy”, the book History of Physics, Desiderius Papp (Espasa-Calpe). The best telescopes of the time came later, from hands of this great master of optics. During a trial with a prism of exceptional clarity,

Fraunhofer stumbled in 1814, with the greatest discovery of his life: the prism had spread sunlight in a broad spectrum that the sage observed through the telescope of a theodolite. Was surprised spectrum vertically crossed by numerous dark lines. Byway, it was not the first to see these enigmatic stripes; two years before him, the English chemist and physicist Hyde Wollaston (1766-1828) the perceived along the bands of the four main colors of the spectrum. He took by the dividing lines that separate a tinge of another and ignored them. Fraunhofer counted more than five hundred stripes, and appointed the most apparent with letters of the alphabet, thus creating the basis of a nomenclature that his successors had nothing but expand. Each stripe, recognized, corresponds precisely determined irrefrangibility. In examining the lines for different positions of the apparatus and different positions of the sun, he saw that those not moving; probably suspected Fraunhofer, are inherent to the same source of light. His interest was from this moment powerfully stimulated: passed the rays of the moon and the planet Venus through a prism. Their spectra appear crossed along the same lines that had

entered the sunlight. The spectra of several stars parading on the screen in the following months; some of these stars, like the Goat, show reproductions, although weaker, solar lines; other patented a different design. Just now, around 1818, arriving from France news on investigations of Fresnel; the obvious interpretation given by the French physicist to diffraction deeply impressed the Bavarian optician, firmly convinced of the correctness of the wave theory. In a series of observations, it replaces the prism glass plates and metal, on which it has drawn very close grooves from each other, up three hundred in one millimeter. These gratings serve to measure the wavelength spectrum dark lines. Admirable is the accuracy of your measurements, whose errors are less than 1 per 1000. Fraunhofer did not just observe the celestial light sources; the flames of the candles and oil lamps will have continuous spectra without limits. It did not escape his attention that the introduction of a salt into the flame to appear in the spectroscope bright stripes and saw the yellow line drawn by sodium flame split into two lines passing through a more powerful prism. The he sought in the solar spectrum and soon realize that the double stripe marked exactly in the same place where they were in the solar spectrum two black stripes, which had intrigued from the day of their discovery and had designated the letter D. sensed the importance of the enigmatic coincidence, unable to interpret. Other research practical absorbed him and served him the mysterious lines as reference signals their search for the refractive indices of different kinds of crystals. His Arcane not disturbed more. As Fresnel, Fraunhofer died of tuberculosis at the same age as the great French: at thirty-nine years. Three decades they had to pass after the death of the discoverer before Bunsen and Kirchhoff were to decipher the riddle of dark Fraunhofer lines and create the magnificent instrument of exploration that is spectral analysis. During that time, many researchers rub the discovery, which always slips from his hands. J. Talbot and Herschel recognize that the same substance that colors the flame of alcohol always emits the same stripes. "When in the spectrum of a flame enunciate Talbot, stripes appear certain and specific, these are safe characteristics of the metal contained in the flame." Wheatstone extends observations studying light of the electric arc, and is different stripes metals used as electrodes. Miller makes solar rays through iodine vapor and bromine to examine the absorption lines. Glimpses of knowledge are born without giving a certainty. Foucault, so skilled in other experiments, this time groping in the dark and does not reach beyond the kind of line D of sodium. Follow the Swedish Angstrom and the Swan, Stokes and Brewster English. The latter recognizes that certain dark lines in the solar spectrum are generated by absorption of the rays in the Earth

Atmosphere. Plus, all these sages only make isolated and inconsistent findings; Finally, in 1859, capital Bunsen and Kirchhoff discovery arises clearly enunciated the law and achieve the first applications. RG chemical Bunsen (1811-1899), as inventive as tireless experimenter, and the brilliant theoretical physics G. Kirchhoff

(1824-1887), both professors of the old and renowned University of Heidelberg, complemented the happiest way and collaboration could not be more fertile. They colored by substances given flames caught the attention of Bunsen, who endeavored to obtain from them a safe means of identifying chemical bodies. Obviously, it was necessary first have a pure flame. The alcohol, with the inevitable impurities introduced by the wick, did not lend; gas lighting seemed more appropriate. Bunsen trials for mixing air with gas explosion lighting without the lighter out in 1884 that bears his name, a constant source of flame, pure, light, auxiliary essential since in laboratories. Bunsen not content to observe with naked eye colors fathered by different substances in the flame of his lighter: the review, following the advice of Kirchhoff, through prisms. The results led him soon to recognize that the bright lines emitted by metal and incandescent vapors are independent of temperature, also independent of the elements with which the metals are combined, and provide safe and consistent characteristics of chemical bodies, although present in small quantities. Suffice less than one ten millionth of gram of sodium to produce the double yellow line that is still indicating the presence of this element when analytical chemistry fails to discover the slightest trace of this. The study of the lines emitted by various bodies, whether in the flame, whether in the arci's in the electric spark, convinced Bunsen safety of his method, soon brilliantly confirmed by the discovery of two new elements. The rubidium and cesium, found by

Bunsen in 1860 and 1861, respectively, received their corresponding names for the spectral lines that allow find. Emission spectral analysis was founded. Required Be completed to become an instrument whose reach once again, as in the days of Newton extends from the Earth to the far reaches of the sky. Transcendental amplification efficiency of spectral analysis meant solving the enigma still open Fraunhofer lines and was the work of Kirchhoff. In the laboratory, artificially Fraunhofer lines in the spectrum was the first and decisive success which gave the key the problem. Kirchhoff and Bunsen executed the feat so that once seems very simple. Kirchhoff ignited an intense flame begetter of a continuous spectrum: in the path of the rays, placed an alcohol lamp with solution of sodium, emitting double yellow line feature. Instantaneously, yellow and bright lines became black lines D, identical to the solar spectrum. If instead of sodium chloride was taking lithium, he saw the characteristic red stripe lithium go dark. He recognized that simply place colored flames, sources of bright lines, between a sufficiently strong light source and the screen and a spectroscope for the flames absorb rays of the same wavelength emitted, and introduced into the spectrum instead, black stripes. "concluded Kirchhoff wrote in October 1859 to the Academy of Berlin- that the dark lines of the solar spectrum that are not produced by the Earth's Atmosphere are caused by the presence in the hot solar atmosphere of substances that in the spectrum of a flame have bright lines in the same place. we can admit that the bright lines the spectrum of

a flame, which coincide with the lines D, are always due to sodium content of these. the dark lines D in the solar spectrum can conclude, therefore, found sodium in the sun's atmosphere. " As gases are solar shell cooler than the sun, a given element of the solar atmosphere is unable to replace its own radiation rays absorbed. Thus, are born the dark lines in the solar spectrum, gaps translate absence in the light beam given elements, and their presence in the sun. The enigma of Fraunhofer lines was therefore resolved, and at same time open the possibility of chemical analysis of the Sun, possibility considered some decades earlier by the French philosopher Auguste Comte like a dream beyond human reach. More here Kirchhoff not stopped; two months after its first submission to the Academy of Berlin, he proceeded to the generalization and rigorous test of the law that he had found. Introduced a new notion, that of perfectly black body susceptible to fully absorb the rays of all wavelengths and not reflect any. Such Body, an integral radiator, there was at that time rather than the imagination of Kirchhoff, and technically was made later, in 1895, by Wien and Lummer. Once defined the black body Kirchhoff proved the validity of equality which is the emissive power, the absorptive power of any body, and E and the powers of emission and absorption of the black body. As it absorbs all rays, A is equal to unity, so that the ratio of the powers of emission and absorption of a given body and / A is a constant determined well. And Kirchhoff stated his law for radiation of the same wavelength, at the same temperature, the relationship between power emission and

absorption power is always the same. The idea turned into reality with the discovery of Kirchhoff and Bunsen, that is given to man penetrate the chemical nature of separated from us by unbridgeable gulfs of space substances, it seemed not only Auguste Comte, the prophet disproved, but to witnesses same feat, incredible and utopian. Fun is read the words of Kirchhoff in a letter written in 1859 to his brother Otto letter: "My attempt, chemical analysis of the Sun, like many very bold I'm not angry with a philosopher at the University for telling me, as we walked, a madman claims have discovered sodium in the sun. I could not resist the temptation to reveal that crazy with Kirchhoff's law, the interpretation of the spectra received a solid base, and the decoding of the spectral signals could initiate, supported a part by the deepening of the emission spectra of chemical elements knowledge, and elsewhere by the growing power of devices Al spectroscope Kirchhoff and Bunsen grating was associated; with the progress of the machine divide, the American physicist Henry Rowland created in 1882 the networks formed by ridges of surprising subtlety to 1100 in a millimeter. Rowland also applied the division grooves concave mirrors. Kirchhoff mapped the solar spectrum, assigning many lines chemical elements that engender. The Swedish AJ Angstrom followed; It was the first to describe the solar rays in terms of wavelength. In the same year, 1868, the English Astronomer G. Huggins led the spectroscope to Syria and applying the measured Doppler shift of the lines, caused by the removal of the star. That first evaluated the radial velocity of a star Kirchhoff.

few months earlier, still in the same year of 1868, a total solar eclipse gave clear evidence of the certainty of discovery of Bunsen and for Few seconds the Sun's photosphere was covered by the moon and suddenly appeared instead of the dark lines, bright lines corresponding lightning spectrum emitted by the solar atmosphere, thanks to eclipse, it was the only shining. A new science was born: astrophysics. In the decades that followed the feat of Kirchhoff and Bunsen, this science put increasing extent, the scope of the physicochemical exploration, not only the sun and stars, but also the spectroscopic eye penetrated inside nebulae, away from Earth for several billion light years. Contrary to expectations, no unknown chemical body in the terrestrial nature drew his stripes on the plates of the English J. Lockyer and attributed first intention to an element that only exist in the Sun, was completed by finding (1895) as forming integral part of the atmosphere of the globe. Spectral analysis revealed the chemical analogy between the stars and rose to rank of certainty the Earth substantial agreement with the remotest stars of the Milky Way and even distant galaxies. The demonstration of the material unity of the cosmos explorable is the sublime lesson, historically the first, we were granted by the spectroscope, thanks to Kirchhoff and Bunsen. However, this success, despite how magnificent, is just one of the many aspects of knowledge opened by the decipherment of spectral lines. They also provide us messages of processes at the atomic mechanism begetter of spectral lines. They are like the distant echo of the configuration changes that are met in the universe of what infiniteness small. Almost all the progress made during the twentieth century in the exploration of the atomic interior, we owe to the depth interpretation of spectral lines. To have extended the scope of the investigation both to the far reaches of the macrocosm and no less fathomless depths of the microcosm is the importance of the work of Kirchhoff and Bunsen, comparable, in its majestic scale, the discoveries of Newton! The beautiful simplicity of the spectra, as demonstrated in the experiences of the two initiators, must soon give way to the realization that the spectrum depends not only on the bodies in presence, but also the way they are excited. The spectrum of a given element changes as may be vaporized in an electric or excited radiation from electric shock arc. A simple flame spectra arc and spark, higher temperature than them, the last studied since 1865 by the Germans Added Julius Plucker and Guillermo Hittorf. Here began a long series of descriptive works exactly for fixing emission spectra of different elements, several of which, such as iron, they revealed its extreme complexity. It is impossible stay here chronicling the laborious and patient investigation that led, thanks to Kayser and Runge, in Bonn, and later Exner and Eder, in Vienna, true encyclopedias of spectral lines. Once measured, after a gigantic work, the spectral lines, and assigned to each element theirs, whether the distribution of the characteristic lines of a given element dispersed over entire length of the spectrum, is not subject to emerged a rhythmic order. One could assume that a certain periodicity was theirs.

A vibrating string a certain number of notes that can covered in a formula in their sounds. The simple formula that theory had established for sound vibrations, would find impossible to light vibrations? The Swiss scholar JJ Balmer (1825-1898) was not the first to propose this, but his rivals had not inexhaustible patience nor shared his unwavering conviction that the law was sought. Balmer, drawing teacher, as artist and scholar, was convinced the omnipresence of harmonious relations in physical phenomena, did not admit that the spectrum could be an exception. His perseverance finally Swiss triumph in 1885, when he stumbled after many calculations, with the numerical relationship between the governing hydrogen lines in the visible part of the spectrum. The empirical formula of Balmer described with extraordinary accuracy the wavelength of the hydrogen lines, where k is a constant "m" can take integer values from three. Kayser and Runge replaced in the law of Balmer wavelength by the frequency and obtained the formula that results in the current notation: where R is a constant an integer greater than 2; each value is a line. The frequencies of the hydrogen lines admirably obey Ballmer's formula. In its discovery, they are also hidden knowledge that the Swiss researcher was far from suspecting. Their discovery soon became a true instrument of prophecy. The formula, widespread in our century by Walter Ritz (1908) allowed envisage not just one, but a whole series of hydrogen lines in the ultraviolet and infrared spectrum. Experience has magnificently justified forecasts and at least in the spectrum of the simplest atoms, hydrogen, chaos yielded to rhythmic order and all stripes together in a formula submitted to the law of 9-Balmer-Ritz. Addition, it was revealed that the lines of other elements also obey similar formulas, even more complex. You are representable by differences of quadratic expressions. The constant R of the law is in the series of spectral lines of all elements; is a universal and fundamental fact, as demonstrated the Swedish physicist Rydberg, whose name was linked with the constant R. By establishing a fixed relationship between emission and absorption of radiation, Kirchhoff opened, as we have just seen, the way to magnificent emergence of spectroscopy; His law clarified many issues, but also gave birth others. Black bodies completely absorbing rays of all wavelengths also emit all, being thus endowed with the maximum emittance. This depends only on the temperature. What is the law of this dependence? The law that links total black body radiation with temperature. Leaning against measurements J. Tyndall and others, the Austrian Physicist J. Stefan (1835-1893) concluded in 1879 that the total black body radiation is proportional to the fourth power of its absolute temperature. After determining the number of calories irradiated in a second per one square centimeter of black body, the Stefan possible calculate the temperature of the Sun, in around figures, at 6000 degrees Celsius, on condition that the Sun is a black body absorbs all radiation, condition that seems, according recent experience, per reality. How is distributed black body radiation on different wavelengths of the spectrum?

This problem already worried Kirchhoff. If a lump of coal or iron is heated with infrared and red rays, which are the first to appear, arise with increasing temperature yellows, blues and violets. The domain of the emitted rays are therefore shifts from low to high frequency. G. Wien studied this relationship, finding in 1894 Bleed law that bears his name: with increasing temperature, the maximum radiation intensity shifts from longer wavelengths to lower, so that the product temperature the absolute wavelength corresponding to maximum is a constant. Stefan's law is an empirical finding; Austrian physicist L. Boltzmann gave the necessary support, to the electromagnetic theory of light, firmly established by Maxwell. More success was soon proved precarious. None of the nineteenth century thinkers turned to the problem of black body radiation could give a satisfactory interpretation of the spectral distribution of energy. While the bell curve characteristic obtained experimentally introduced a maximum whose position was regulated by the law indicated Wien, the theory demanded a curve whose coordinates grow to infinity, when the wavelength increases. Nature reveals once again that their laws not always accommodate the reasoning of our spirit. Only the twentieth century physics frees quagmire that led to the patent contradiction between theory and experience. With the new century born the new doctrine; the December 14, 1900 suggests Max Planck (1858-1947) the innovative idea of considering the radiant emission as a batch process that is carried out by isolated elements of energy, holders of a certain magnitude. Such Element, as is proportional to the frequency of lightning, the proportionality factor being a universal constant of nature, the famous constant h which was later immortalize the name of its discoverer. Thus, the energy of a is as given by the formula. The lucidity of this thought suddenly clarified the enigma of black body radiation immediately explaining the variation of the bell curve, whose caprices had puzzled researchers. Such success was but the first achievement of the new theory. Assuming the fertile germ Planck most unprecedented and wonderful ideas that should transform beyond recognition, the image of the physical world was hiding. One of his most resounding victories should be the explanation of spectral lines by changing the electron interatomic configuration. Balmer law Ritz admirably described the hydrogen lines, but nothing reveals why a certain line radiating element and not another; He left completely in shadow the mysterious bond between radiation with the atom radiator. Only when the Danish sharp Niels Bohr (1913) introduced the quantum atomic interior, giving the circulating electron trajectories governed by Planck's constant as assumed, with a bold hypothesis that the electron emits light to jump from one orbit to the other, succeeded obtaining from the energy differences orbits frequency of the emitted radiation. As by magic they appeared in calculating the frequencies of spectral lines. But now the heroic age of spectroscopy had long belonged to past thalamus.

The first gland sodium and potassium creation process to create proteins as mentioned in previous paragraphs. Allegre hit the spot with Bunsen and Kirchhoff observations. The only mystery now I could recognize and bind him to the theory proposed in my first book, By the feather of the bird and dry flower. The sun emits energy radiating towards the particles Earth sodium and potassium, basic elements of creation, together with phosphorus, sulfur, carbon, oxygen, hydrogen, salts and oils from the composition of other elements that come in the game of evolution. These are elements that are concentrated in the waters of lakes and seas, and contained in the earth itself. The first stage of evolution of being is based on the interaction of these elements and an energy that bounced off the sodium and potassium elements, and activated or the spark of life on planet Earth.is possible that in the universe operate the same but nature created and manufactured with materials and elements aggregation meet needs. It is also for the aggregation of materials that inherited qualities are created and then manifest in the thing created. It is a quality of the building materials adding to existing and generate new forms and energies. Oogenesis is currently studying a phenomenon that looks like this stage of early development pattern. Possibly, as in other species, humans had the ability change sex in their original stages. It is a quality of some species that existed for millions of years.

The simple sex change or reversal, removal of one another, was a quality that became man in the stages of incarnation, where the quality asleep during a stage of life arises in the other and vice versa when being regenerated in a new body. The proof is that in area of the clitoris is that female hormones create male or female member by selection. My personal analysis Based on the findings and information provided in these documents comes an observation about recent discoveries around California, where a cell is created and evolves with arsenic. It would be normal to be developed with potassium or sodium, as in the first cell that came with the qualities of the materials that survive will equip humans the origin we fear. From the observations, I have made on these findings it shows that their arsenate potassium and sodium arsenate, since it has denounced the coasts of Mexico and other countries, the presence of these compounds. If by some chance has developed a primary cell of this compound, it would be another example of evolution, such as the thalamus in humans. But I worry that the observations only They point to a single element in this interaction and that does not comply with the law division of creation, because must be a compound, which gives rise to the emanations of life. I understand a simple law of creation and is that two conditions must be present for the emergence third. In metaphysics, it is the law of the triangle of manifestation, therefore, microbes, plants and animals, can convert all these chemical compounds of inorganic arsenic in organic compounds, comprising carbon and hydrogen atoms. The spark of energy one, the sodium and potassium, which act with the first radiation energy. Is a new element that provides a recent test to be discovered about a year ago, and its effects have not been disclosed to present, are evidence that salt water can recreate intelligent life. May be an isolated phenomenon that is disclosed now, but God knows how long this is happening. Interesting would be check whether some form of intelligent life has emerged in this coast, that relates to this phenomenon and can be checked in a future flash in ways that give light to know about. Compounds: Techniques chemical analysis Chemical Laboratory Training College of Bachelors of the State of San Luis Potosi. 28. campus sodium arsenate potassium

Introduction.

In several Latin American countries like Argentina, Chile, Mexico and El Salvador, at least four million people drink water permanently with arsenic levels that jeopardize their Health. Arsenic concentrations in water, especially groundwater, present levels reaching in some cases up1 mg / L. In other regions of the world, such as India, China and Taiwan, the problem is even greater. By information obtained, in India there about 6 million people exposed, of which more than 2 million are children. In the United States, more than 350,000 people drink water containing arsenic is greater than 0.5 mg / L, and more than 2.5 million people are being supplied with water with arsenic tenors greater than 0.025 mg/ L. The problem of arsenic in drinking water has been treated in Argentina for several years, when epidemiological Cordoba and other provinces of the country showed and associated disease Making,

damage to the skin, the presence of arsenic in drinking water. Efforts and studies conducted to minimize or eliminate have made a breakthrough at level of water treatment urban scale in Argentina, Chile and Peru, but, at rural level, the solution in these countries still pending. Hence Argentine health authorities decisively promote studies involving a proposal for the solution or minimization of the problem identified. Effects of arsenic in humans is known that the main routes of exposure to arsenic are ingestion and inhalation, which is accumulated in the body due chronic exposure and certain concentrations causes problems such as skin disorders, relaxation of the capillaries skin and dilation of these, with side effects on the nervous system; irritation of the organs of the respiratory, gastrointestinal and hematopoietic system; and accumulation in bones, muscles and skin, to lesser degree in liver and kidneys. There Epidemiological evidence that people with prolonged ingestion of inorganic arsenic via drinking water, have plantar hyperkeratosis span, whose main manifestation is the skin pigmentation and calluses on the palms of the hands and feet. Arsenic in natural water arsenic in surface and groundwater. Arsenic occurs naturally in sedimentary rocks and volcanic rocks, and geothermal waters. Arsenic is found in nature most often as arsenic sulfide and arsenopyrite, which are found as impurities in mineral deposits, or as arsenate in surface and groundwater. Arsenic is commercially and industrially used as an agent in the manufacture of transistors, lasers and semiconductor, as well as in the manufacture of glass, pigments, textiles, paper, metal adhesives, food preservatives and wood, ammunition, processes tan, pesticides and pharmaceuticals. Arsenic is present in the water by natural dissolution of minerals from geological deposits, discharge of industrial wastes and atmospheric deposition. In surface waters with high oxygen content, the most common species is arsenic oxidation state +5 (As + 5). Under conditions generally reduction in lake sediments or groundwater arsenic predominates +3 oxidation state (As + 3), but can also be the ace +5. However, the conversion of as + 3 to as +5 or vice versa is quite slow. The reduced compounds of AS + 3 can be found in oxidized and the oxidized compounds means of A s + 5, in small media. Microbes, plants and animals can convert all these chemical compounds of inorganic arsenic in organic compounds -compromised carbon and hydrogen atoms

90-"The creation began and evolved in salt water"

Conclusion NO 1

Therefore, under the new findings of science and its great exponents is possible say that a new link of human knowledge can be assumed as true.

1 The writings of Dr. Esther Rosón Gómez genetics of the brain and its functions, the interaction of sodium and potassium as main sources that him gave the ability to the first cell to produce the proteins and create the first membrane of creation to begin evolution and other functions reveals. Which only exist as elements and be present for millions of years of existence, no purpose defined as the most accumulated elements in outer space and on Earth. The moment that arises a new physical action of these elements could be affected and penetrated by a foreign phenomenon that was not present before, that triggered the energy that started this new manifestation. In some unknown region, exudes a subtle energy composition altered the movement of electrons of the primary field to start creating.

2. Claude Allegre, in a little science for everyone, alerts us to the discovery of Bunsen and Kirchhoff astrophysics chemical and advances refractions of sunlight, where we confirm that sodium and potassium are product of solar radiation and that our planet is penetrated by these and other elements emanating from entire cosmos, saturating the earth and our seas from primary era. And, moreover, all scientific theories promote our system of internal energies, from Newton to Albert Einstein.

3. Studies of the College of Chemistry Institute in San Luis Potosi in Mexico on arsenic compounds, especially the elements derived from arsenate potassium and sodium arsenate, which is coupled with the findings of Dr. Esther Rosón Gomez, who can be a source for future science research.

4. The new finding NASA on a cell using arsenic instead of phosphorus to reproduce. Perhaps only they got a link; there must be many outsides of human knowledge that have yet cataloged. It is like looking orchid in the swamp.

Discovery NASA

A strange bacterium can survive without one of the fundamental building blocks of biology. A bacterium found in the arsenic rich waters in a lake in California is expected to give a turn to the scientific understanding of the biochemistry of living organisms. The microbes seem can replace the phosphorus by arsenic in some of their basic cellular processes, suggesting the possibility of a very different biochemical to hitherto known, which could be used by organisms in past and present extreme environments Earth, or even other planets. Scientists have long considered that all living things need phosphorus to function, along with other elements such as hydrogen, oxygen, carbon, nitrogen and sulfur. The phosphate ion, PO43-, plays several pivotal roles in cells: maintains the structure of DNA and RNA, combined with lipids to create cell membranes and transports energy within the cell through the adenosine triphosphate molecule (ATP). But Felisha Wolfe-Simon, geo microbiologist and research fellow at NASA astrobiology, based in the US Geological Survey in Menlo Park, California, and colleagues online report today in the journal "Science" that a member of the family Proteus bacteria can use arsenic instead of phosphorus. The finding implies that "potentially can be omitted match the list of required elements for life" says

David Valentine, geomi crobiologist at the University of California, Santa Barbara. Many science fiction writers have proposed life forms that use alternative basic blocks, often silicon instead of carbon, but this is the first case of a real body. Arsenic is positioned just below phosphorus in the periodic table, and the two elements can play a similar role in chemical reactions. For example, the arsenate ion, TEP 43- has the same tetrahedral structure and binding sites that phosphate. It is so similar that can enter cells supplanting the phosphate transport mechanism

5 - studies Kristein Dr. M. Kneeling on the influences of electromagnetic energies of the planets on human behavior, and its conclusion this. it was something of which the ancients had control. Its effects benefit of knowing these scientific data and apply personally. The electromagnetic fields of the earth exchanged for different periods of time. Currently the change began in 2005 and continues affect the electromagnetic orientation of the earth. She proved that the earth is naturally aligned to the changes. The same true with animals. The only one who does not have that natural ability is the human being. The person being manager of the functions of the astral and physical body must put on alert as understand its operation and update for benefit of present generations. Function belongs to health agencies to guide and raise public awareness of how handle these data directly affect the individual citizen, the way to act to harmonize their health with the correct data. The attitudes of the masses in different countries

are bound to suffer from possibly catastrophic influences these imbalances in physical laws. It draws on the disorientation and the current violence to these variations. This was a task of NASA to study the "Dead Sea Scrolls" and finding surprised with the unusual discovery, a practice used the inhabitants of those times to use a concentration exercise to balance the natural forces of being with such deviations of the matter. If the position of the Egyptian statues with the usual posture of feet apart and hands on thighs note, that was a mystical posture had intended maintain a balance of internal forces with energy universal. These new findings credited to the rudimentary statement in this book and the theory presented in it. the purpose of exposing more elements that collect information conducive to support this theory suggests me that irrelevant information already exists and has never been linked with this phenomenon of primary evolution. If we take the gland thalamus as the source of divine creation of human beings, we must turn our attention because that there is found the energy to direct to electromagnetic fields and achieve a glandular balance and hormone in our body

6. the origin of being in the saltwater Dr. Linus Pauling Along with René Quinton, that before they presented conclusions on the same subject in 1904. Linus, he postulated that human beings have in body 118 elements in the periodic table, which are present in the salt water bodies such as lakes and seas. Also, announced that the cell life originates in the Precambrian 3,800 million years ago, in salt water. Another finding is that human plasma is like seawater into its components. With remarkable variation that humans now consume fresh water and should have a small variation in composition, but consume processed salt

Conclusion: No- 2 are 6 tests and additional data that exert a powerful judgment on the statements set where the be taken as true in its details, is part of that separate test of the same conclusions, "humans originated in salt water." Gaspar (Edwin) Pagan

The human brain

seething energy wave frequencies and rates of vibration in the minimum and maximum scales that being can imagine. The aura in humans is the emanation of the brain of energy and these transcend their interior to the outside, where produce a halo of light that covers the seven colors or white light refractory cosmic spectrum. In elementary school, he went in the morning to a small kiosk to buy candy. Daily watched a huge pine that was next and felt that at the top, looked like the tip of an umbrella, light beams gave off silver color. It was like a

flow of energy that followed the dome where the wind wants move. A strange sense of vertigo through my body to look at the heights as happens to many people when they are in a tall building and look down or up. I was amazed by this phenomenon. I never told anyone, not knowing what was. After so many years, I could understand that he was facing one of the mysteries of nature: creation. Scholars and mystics scientists are wound brains trying to the explanation of everything that is recorded and which refers to mystical knowledge hidden in symbols that others skilled in the art have accumulated for years around the planet, hiding veiled secrets and hidden meanings. This was the act of combining energies of the forces of the cosmos, Earth, serving as intermediary shaft, which in turn retained part of this for growth and development, and subtle forces released into space from Earth. The same forces of electric shock or lightning, do not have the subtlety of the pine. I also learned from this interaction must have created a need to attract these energies for life is revealed. This is a conscious or unconscious need be. Which achieves decipher its teaching and apply will be in harmony with God Creation: The Holy of Holies. Possibly are the same energies flowing through the dome of the pyramids of Egypt and penetrated to the inner chambers of these, and should have manifestation of this energy that was the secret knowledge that motivated the Egyptians to build many monuments. A demonstrable fact in the teachings of Dr. Linus Pauling that relate the findings in my statements are related to the energy used the Egyptians and other races to bring its interior with the cosmic forces. The thing created in any of the kingdoms manifested establishes a harmony with the divine mind. It is the universal mind that everything contains the logos of creation, because emanates from a universal active mind where all content is.

Being consciously discover the channel of communication with this creative source will be an instrument of creation, a channel that to attract and emanating source, such as pine, energy that offsets the law of creation. Up and be part of it by time intervals and regenerated back to this plane of manifestation. After meditating on all these features contained in this paper and if in them any scientific logic is contained, it can be argued that any external or internal phenomenon to change the laws to which followed this evolution a phenomenon that could disrupt life either complete or partial. That said, under the reasoning that any phenomenon that alters the harmonic confluence with what has already been developed in this physical plane obeying these laws is logical think that these same subtle laws of creation can be altered by events outside the common manifestations of physical laws to which we are exposed. Internal systems of the realms of creation or evolution overcome by overcoming internal conditions and external influences, which the body produces and adds to there is a need to process the advancement or change vital for internal harmony structures. Any variation that is a violent change their DNA structures breaks the laws of evolution, therefore, his own life. The characteristics of anything created succumb to illogical change. For example, just because a plant change to a different climate makes it to wither and die, because he has led to a place that acquires the energies are not harmonious with its development. Otherwise if you are in a poor environment and change to a more harmonious development with the way it acquires a vitality that would further growth and vitality. Not only physical phenomena like fire, violence of physical phenomena, can only partially destroy one thing, but then if a seed vestige rise again this life is. With only address a thought as simple as this opens a way to speculate on the reasons that can affect evolution. It may be a genetic phenomenon that manifests itself after a natural phenomenon.

In the demise of the dinosaurs and other species can be argued that they could undergo genetic changes in their DNA or brain condition be affected by the blockage of energy for body evolution, when seeing their bodies slowly reduced by blocking substances necessary for life continue. The possibility that their hereditary genes override the ability to reproduce or their eggs fertilized. The blockade of cosmic emanations of energy needed for your brain supply the amount of energy to the composition of their huge bodies and these deteriorate and disappear by genetic deletion. The dark corners of genetics and evolution have many gaps, as both creates and destroys, the reason God only knows. The fact that so many phenomena are manifested in changes in nature, the emergence of new species and the disappearance of others should lead us to reason in this way. Corals banks and other fragile species disappears slowly, study its physical causes to discover why.

I suggest in the next chapter possible incidents that happens for the disappearance of species.

Part II

97-Breaking Myths

Pangea

Detachment continents

Uruguayan archeologists have found fossil remains of nearly 130 million years in the fossil deposits of South American country.

This the oldest dating grant that the remains of the dinosaurs that populated the planet millions of years ago, this is said to belong to the Gondwana, the southern continental bloc that emerged after the first separation of Pangea.

The second important in the breakup of Pangea phase began in the Early Cretaceous (150-140 Ma), when the minor supercontinent Gondwana it separated on several continents (Africa, South America, India, Antarctica and Australia). About 200 Ma, the continent of Cimmeria, as mentioned above (see "The formation of Pangaea"), he collided with Eurasia. However, a subduction zone was forming it as soon as Cimmeria collided.
[22]

They have emerged archaeological excavations evidence showing history data of Dinosaurs on the continent of South America. Pangea is the closest to the disappearance of these large animal's phenomenon.

Supercontinent that existed at end of the Paleozoic and early Mesozoic grouping most of the landmasses the planet. It was formed by the movement of tectonic plates, some 300 million years ago, all previous continents joined in one; later, some 200 million years ago, he began to fracture and disintegrate until reaching the current situation of the continents, in a process that continues. This name was apparently first used by the German Alfred Wegener, lead author of the theory of continental drift in 1912.

PANGEA

The first continent that broke in segregated worlds, a debacle at level of our planet after millions of years of existence, I believe consequences that humans still was unable decrypt entirely.

Pangea- Comes from the Greek "pan" prefix meaning "all" and the Greek word "Gea"
"soil or land"

Uruguay

Found in Uruguay fossils older than the dinosaurs, mean that is an earlier event at the age of dinosaurs.

the Uruguayan archeologists have found fossil remains of nearly 130 million years in the fossil deposits the South American country. This dating gives them greater antiquity than the remains of the dinosaurs that populated the planet millions of years ago, this is said to belong to the Gondwana, the continental block south that emerged after the first separation of Pangea.

The latter finding is a species that has two rami articulated, in which each has the trigeminal nerve and capillaries. Paleontologist, Piñeiro explains: "We have not found a single reference that has previously registered a similar peculiarity preservation. This feature is what makes the discovery is so No fossil record has similar remains. For paleontologist, the new finding means to go further. With new discoveries, experts will "go a step further" and meet animals that lived nearly 300 million years ago in a hostile environment: "What we discovered enables us know aspects of behavior that are little fosilizables. That is, how eat, how they reproduce, how they adapt so well to an environment that is not very favorable to life, as it is a very salty and low oxygen "lake.

The place where they found the remains fossil known as "Konservat Lagerstätte". Are fossil beds where, under very specific conditions, structures that normally would not preserved elsewhere. The fossils found by the research team were in a group of rocks over 280 million years called "Mangrullo" ago. This Cover area of Tacuarembó, Rivera and Cerro Largo, although the rocks extend to Brazil. Piñeiro said: "We do excavation work on Mangrullo rocks on the surface. These rocks continue in Brazil and we work together with geologists and paleontologists from that country "move north. In the Cretaceous, Atlantic, South America and Africa today, finally separated from eastern Gondwana (Antarctica, India and Australia), making the opening of a "southern Indian Ocean". In the Middle Cretaceous, Gondwana fragmented to open the South Atlantic Ocean, South America began to move westward away from Africa. The South Atlantic did not develop uniformly; rather, he rose from south to north.

addition, at same time, Madagascar and India began to separate from Antarctica and moved northward, opening the Ocean.
Indian Madagascar and India separated from each other 100-90 Ma in the Late Cretaceous. India continued move northward toward Eurasia at 15 centimeters (6 inches) per year (a record of tectonic plates), closing the Tethys Ocean, while Madagascar stopped and turned to the African plate blocked. New Zealand, New Caledonia and the rest of Zealand began to separate from Australia, moving eastward towards the Pacific and opening the Coral Sea and the Tasman Sea.

101- ARDI oldest human 4.4 million

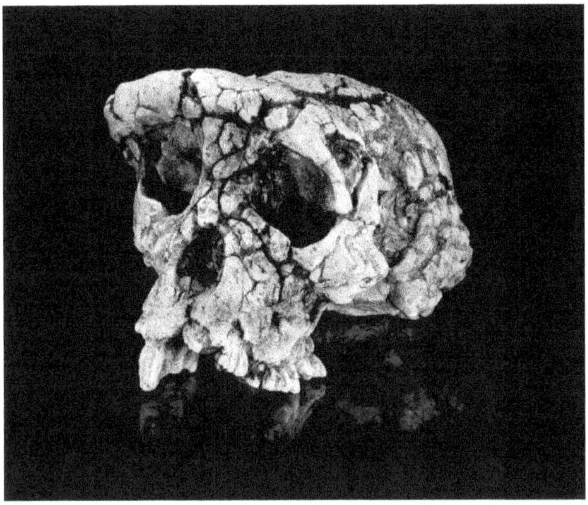

Finding 4.4 million years' specimen.
What is now Ethiopia demolishing the theory of Darwin. It is proof that from Pangea and the time of the dinosaurs and lived humans on the planet evolving fought adversity and survived.

After the recent discovery, the site of Uruguay is the oldest in South America and the world's second largest. The reconstruction of the place where the animals lived 280 million years, is funded by the National Agency for Research and Innovation (ANII) of Uruguay and the results will be published in various scientific journals around the world. This description gives us a clear understanding of the phenomena that have hit the planet earth.

Our mind and history is written by those who have the power, if they seek the date on which the figure of Darwin and his theory arises, we must bind civilization and the mastery of this by interested created. Search in the basement of the creations of the advanced minds, is a maze for common being

that has the advantage of having to create extraordinary stories, before the cinema or television fortunes. Fantasies that occur today take the same course, create truths disguised for common being that can't penetrate the truth.
A child begins life most through cartoon train him to believe anything that makes a piece of cardboard, or drawing. The films have become famous in that sense, everything is possible be created in the mind of humanity, because he has trained us to believe anything. No one questions the application of unnecessary details while your mind is occupied. That humanity has achieved with this practice. Every day being further from the truth and the powerful can manipulate human intelligence.
This finding puts an end to Darwin's theory and the idea that man descended from monkeys. In fact, the remains of Ardi offer a window into what the last common ancestor of humans and apes living today could have been a mother with all the attributes of the modern woman. In different territories are facts that bound in information opens a new route to the science of the future to clarify taboos.
In this they continue the findings in the area and have found many fossils more, both contemporary Ardi and Lucy as well as older ones, such as "Chad" whose remains date from approximately 6,000,000 years. An international team of scientists presented they say is the oldest and best preserved of a direct ancestor of the human species fossil.
Ardi, catalog name ARA-VP-6/500, is the nickname given to the skeleton of a female belonging to the species Ardipithecus ramidus, probably a hominin (bipedal primate), which is considered the most primitive hominin known to date and he lived during the Pliocene, about 4.4 million years.
An international team of scientists sat the pre- they say is the oldest and best preserved of a direct

ancestor of the human species fossil. This is a female spice Ardipithecus ramidus, which lived 4.4 million years ago, in what is now Ethiopia.

As the researchers note in the journal Science, if it were not our direct ancestor, the finding provides valuable information about a crucial stage in human evolution: the time when we separate from the common branch that we share with monkeys. The discovery, say the researchers, shows like never biology of this first stage of human evolution.

Until now, the oldest known stage of human evolution was Australopithecus, the small-brained biped. It is older than the famous Lucy of Australopithecus, a fossil of 3.2 million years discovered in 1974 about 70 kilometers from where it was found Ardi.

When Lucy was found, the international community thought that the earliest hominids would have a similar anatomy of chimpanzees, but Ardi, which is almost one million years older than Lucy, does not support that theory. It is good-looking? The found in Ethiopia, it is a female of the species of hominids. The fossil (Ardipithecus ramidus) tops the list of the most important scientific advances and is 1 million years older than previously found last on the origin of man. 'Ardi' has more than one million years old Lucy (Australopithecus afarensis), a partial skeleton of what was considered until now the oldest hominid. The investigation of Ardipithecus has changed "the way we think about human evolution and represents the culmination of an arduous research collaboration of 47 scientists from nine countries analyzed 150,000 specimens of fossilized animals and plants," said Bruce Alberts, editor Science in an editorial.

After analyzing the skull, teeth, pelvis, hands, feet and other bones of the fossil, the scientists determined that "Ardi" had a mix of "primitive" features shared with apes of the Miocene who preceded him and other features "derivative" which shares exclusively with hominids. The discovery of the fossils made in 1994 revealed the biology of the early stages of the better than any other to date human evolution, said the American geologist Giday Wolde Gabriel, who led the analysis of lavas and ashes that were used to determine the age of his remains. The set of fossilized bones of "Ardi" revealed that it was a woman who weighed about 50 kilograms and had a height of 1.20 meters. "With such a complete skeleton, and with so many individuals of the same species in the same time horizon, we understand the biology of this hominid," said Gen Suwa, paleo-anthropologist at the University of Tokyo and author of a report published by Science in October this year.

Ardi

The fossil reveals that the ancestor of humans suffered a stage of evolution poorly known a million years before Lucy (Australopithecus afarensis), the iconic fossil female who lived 3.2 million years ago, and was discovered in 1974, to just 74 kilometers from where Ardi was found. The researchers argue that the shape of the pelvis, members suggest that it was bipedal when walking on the floor, but it was quadrupedal when moving among the branches of trees.

Ardi was partially specialized in the upright posture, with the pelvic girdle somewhat "acuencada" to support the intestines in a vertical position and with the bones of the feet slightly rigid to facilitate the biped displacement.

While she had long arms and curved fingers to grip the branches of trees, also had in the foot, divergent from the other fingers, as in the great ape's big toe, or hallux, allowing him to grab the foot, also the trees.

Ardi, as it has been dubbed, was discovered in 1992 in the Afar region in Ethiopia, but it took 17 years to carry out the analysis of the find.
The fossil has a small brain of 300 cm3, he belonged to a female and was dubbed "Ardi". The radiometric dating of the layers of volcanic lava show that Ardi lived 4.4 million years.
Species: Ardipithecus ramidus
Gender: Female.
Weight: 50 kilograms.
Height: 120 cm.
Preserved remains: hands, feet, legs, ankles, pelvis and most of the skull.

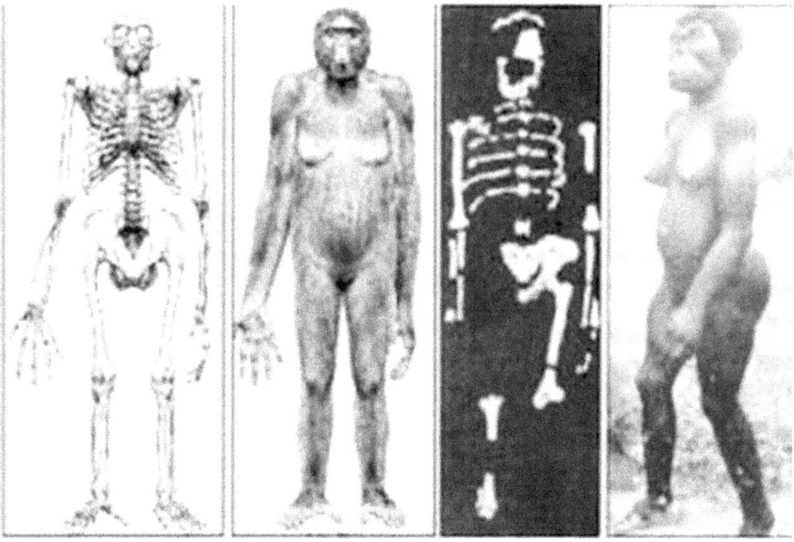

3- Bulgaria

Sofia News Agency, Novinite reports that the skeleton was found in what was Giant human skeleton unearthed in Varna,
Archaeologists in Bulgaria have discovered the remains of what they have described as a "huge skeleton" in the center of Varna, a city on the Black Sea, whose rich culture and civilizations extends for about 7,000 years. The ancient city of Odessus, a trading post established by the Greeks towards the end of the seventh century BC. Odessus was a mixed community of Ionian Greeks and the Thracian tribes (getas, Krobyzoi, Terizi). Later it was controlled by the Thracians, Macedonians, then the Romans. The Roman city, Odessus, covered 47 hectares in the current central Varna and had prominent public baths, Thermae, built in the late second century AD, now the largest Roman rest in Bulgaria.

The Nephilim Chronicles: Fallen Angels in the Ohio Valley
Did a race of giant humans once roam the Biblical lands, Europe and North America? Over 300 historical accounts of giant human skeletons are presented for the first time. Massive human skeletal remains, burial mound types, symbolism, etymology, numerology and ceremonial centers are compared in the Biblical Levant, the British Isles and the Ohio Valley with stunning similarities. Genesis 6:4, "There were giants in the earth in those days; and after that, when the sons of God came unto the daughters of men, and they bear children to them, the same became mighty men which were of old men of renown." The giant offspring of this union between the sons of God and the daughters of men were called the Nephilim....Frit Zimmerman

Read more: http://www.ancient-origins

Fallen angels in the Ohio Valleyorigins.net/newshistory-archaeology/gianthuman-skeletonunearthed-varna-bulgaria002787#ixzz3W6WvVVmb
Follow us: @ancientorigins on Twitter
Ancientoriginsweb on Facebook

It is not the first time a large skeleton has been found in Eastern Europe. In 2013, the skeleton of a giant warrior dating from 1600 BC was found in Santa Mare, Romania. Nicknamed 'Goliath', the warrior is more than 2 meters high, very unusual for the time and place, when people were of small stature (about
1.5 meters on average).

4-Human Lake Delavan in Wisconsin.

The warrior was buried with an impressive dagger indicating its tall.

The news had a great echo and caused a stir, so much so that the New York Times reported the news on its pages. Perhaps, in those days, there was more freedom and less fear of discoveries that can change the well-established scientific beliefs based only on theories. So, writes columnist New York Times article published May 4, 1912.

"The discovery of several human skeletons while excavate a hill in the Delavan Lake indicates that a race of men hitherto unknown once I live in southern Wisconsin. [...]. The heads of these men presumably, are much larger than the heads of any race that inhabits America today.

The skull seems to stretch back immediately above the eye sockets and nose bones protruding well above the cheekbones. The jaws appear to be long and pointed

MISSING SKELETONS OF ANCIENT RACE OF GIANTS THAT RULED America

CATEGORY: CIVILIZATIONS

12/25/2013
Read more: http://www.ancient- origins.net/news-history-archaeology/giant human-
skeleton-unearthed-
varnabulgaria002787#ixzz3W6XDBE9E

There are discoveries that, for reasons not entirely clear, is stored in the oblivion of human knowledge. These findings may shed light on the distant past of mankind, however, they are shrouded in fog and with many contradictory time lines.

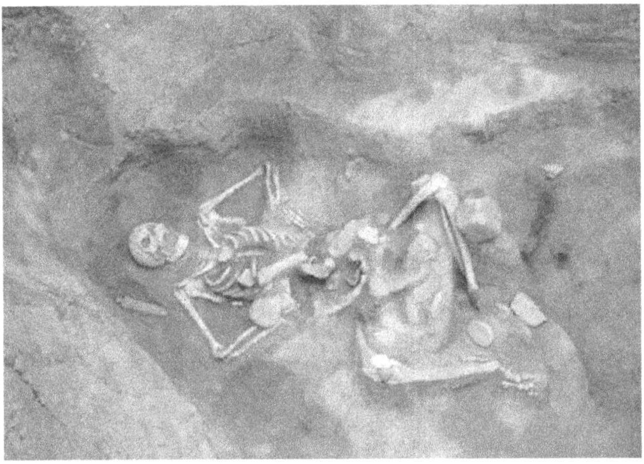

In the case of North America, there is evidence that these skeletons were destroyed to not reveal the origin of races that inhabited the continent. In countries like India, Egypt and other no physical evidence in museums.

In May 1912, a team of archaeologists from the Beloit College in the US, in an excavation on Lake Delavan in Wisconsin, brought to life more than two hundred effigy mounds were considered - as a classic example of culture Woodland, a culture that American prehistoric dating back to the first millennium BC believed.

New York Times, 1912

The description of the skulls provided by the New York Times, reminiscent of the shape of the skeletons belong to the recently discovered in an ancient burial in Mexico with the difference that we are dealing with individuals taller than three meters. Who were they, and why there is no trace in the official history we were taught in school?

Are these human giants lived on our planet, and in any case, belong to humans? Can this be an ancient settlement of ancient humans, survivors of the tragedy of Atlantis? Or were beings from other worlds that corroborates the theory of ancient astronauts? It is hard to say.

Alternative Archaeology (AIAA) that the Smithsonian Institute had destroyed thousands of remains of giant human during the early 1900s was not taken lightly by him, the accusations stem from the American Smithsonian Institution responding suing the organization defamation and trying to damage the reputation of the institution 168 years old.

Conclusion

There is no accurate appreciation of creation, neither science nor theology are branches of confidence for the understanding of life, have always played dice with the universe, humans are puppets drivers, being must be the main actor on this stage creation, take control of their powers and establish their own personal world, not let gods play with your life, we have the freedom to choose, a direct divine gift to man. The information is released to the public and does not flow as it should be as this not seek prepare your intellect.

Personal disclosure based on the facts accumulated

The man finds God behind every door that science can open.

Albert Einstein (1879-1955) German scientist naturalized US citizen.)

These data against science with the origin of creation that surrounds us in the theories presented in this document as proof that the origin of beings and species in the oceans is based with the first evolutionary appearance of beings on the earth. As Clement of Rome argues in its concepts of mystical philosophy on life of those who contemplate later development at the time of the life of Jesus, the example came on the form of the divine consciousness manifest in beings, in the human realm. In other realms happens the same, but has not delved into the perfection of innate intelligence of these creations is the most imperfect humans.

111-The beginning of life

A be initiated into the mysteries unveil the of creation hidden in the consciousness of God, his eternal and hidden wisdom. This is the path of science personnel that scrutinize the universe and give us their findings with an open communication of the truth.

Details took in the sight of intuition that universal mind. The cosmic archive of human existence in relation to the plane of inner maturity, the continuous flow of its essence into being since its appearance on this plane of creation.

The human mind is a receiving channel between creation and creation is the conscious mirror of the universal mind and its processes that can be harmonized with due knowledge. Ambiguous concepts of tradition on creating myths about adopting processes lead to active to deviate from the concern deepen the knowledge mint. It is assured that someone controls everything and so you need not worry or go beyond what is proclaimed.

Through the centuries have been devoted minds that have transcended the sublime creation, his legacy is what is studied in history, and the face that has not been revealed occult, which belongs to a dimension of control. The active mind can access incarnations and that knowledge has done with proper guidance and preparation.

The spiritual legacy of many embodied beings, we have brought to humanity at different times, the sublime messages to raise awareness of the other brothers of creation. The common mind is subject to convictions created by free choice, we are a sponge everything that is projected and touch our perception, being discerns and accepts what his mind reasons per their maturity. Staged uses that knowledge as a point or focus of orientation, the truth of creation is a complex outer and inner world for being, its stages projection caused a temporary impression of reality.

The universal laws are an expanding spiral, its truths are cosmic creations that the mind completely cover, progress data will be added to the reality of each being staged. The advancement of functional reality depends on the advancement and revelation of the eternal truths each scale is revealed a different world, a variable print for each individual consciousness and becomes collective through communication. The data released most of the time fall on vested interests that are not willing to let the truth flow freely. This is the dam that society must face to discern the truth of what is projected, open your individual consciousness and see the reality of life, free-thinking and creating.

We must raise our minds to cosmic and physical phenomena, because the creative energies emanating from those laws. The interior structures that do not adapt to genetic changes due to their structures and forms of change and internal adaptation. Cells have saved a genetic code and do not adapt to any changes, it would be another link in the very nature of evolution. Laws to obey all species in all realms of creation are controlled by the emanations invading our physical and internal environment of the energies that interact with our physical laws. What it is called emotional because they are affected by the internal harmony of the conditions that establish a sense of well-being, internal and emotional maintaining an existential balance. If we seek a reason for the disappearance of the dinosaurs in the forests of the land, due to the phenomena of asteroids and fumes, a phenomenon that has destroyed the existence of all animals, plants, humans, seeds, eggs, worms. Possibly the effects of the devastation of energy, by block entry into the atmosphere affect glandular systems and structures of the bodies depends on these, therefore destruction being the lack of energetic material for the internal laws of growth and making the necessary elements for growth deteriorate, the level that most of life is altered sequence and stop manifested by a relative period. With the publication of Pangea where they have found as many dinosaur fossils, it warns us that some phenomenon occurred in that

113-THE ESSENE GOSPEL OF PEACE

The way in which the truth was concealed from humanity to keep her locked in ignorance, control of wisdom at the expense of the lives of the daring who sought to reveal the truth, sword and fire singed flesh was the bravery price. They paid the price, gold seized its assets.

1- And then many sick and maimed came to Jesus, asking, "If all you know, tell us why we suffer these painful pests Why are we not whole like other men Master, heal us, to give us strong and not? we have to live longer in our suffering. we know that your power is to cure all kinds of disease. Deliver us from Satan and all his great evils. Master, have mercy on us. " And Jesus said, "Blessed are you who hunger for truth, for I shall satisfy with bread of wisdom Happy you who call, then you'll open the door of life Happy you who reject the power of Satan, then I will lead. the realm of the angels of our Mother, where the power of Satan cannot penetrate and they asked in bewilderment: "Who is our Mother and what are the angels?? And where his kingdom is "" Your Mother is in you; and you in it. She gave birth to you and she gives your life. It was she who gave your body, and it return her again someday. Happy are you when you come to know her and her kingdom; if you receive to your Mother's angels and are fulfilling their laws. Truly I tell you, whoever does this will never know the disease. For the power of our Mother she is above all. And it destroys Satan and his kingdom, and has rule over all your bodies and all living things. "The blood that we run is born of the blood of our Earthly Mother. Her blood falls from the clouds, springs from the bowels of the earth, mutters in mountain streams, flowing spaciously in the rivers of the plains, sleeping in lakes and mighty rages in stormy seas. "the air we breathe is born of the breath of our Earthly Mother. Her breath is azure in the heights of heaven, whistling in the tops of the mountains, whispers among the leaves of the forest, waves over the wheat fields, slumbers in the deep valleys and scorches in the wilderness. "The hardness of our bones is born of the bones of our Earthly Mother, of rocks and stones. They stand naked to the heavens on top of the mountains, are as giants that lie dormant in the foothills of the mountains, as idols raised in the desert, and are hidden in the depths of the earth "the delicacy of our flesh is born of the flesh of our Earthly Mother; meat ripe yellow and red in the fruits of the trees and feeds us in the furrows of the fields. "Our bowels are born of the bowels of our Earthly Mother, and are hidden from our eyes like the invisible depths of the earth. 5" The light of our eyes and hearing of our ears are born both colors and sounds our Earthly Mother, that envelops us like waves of the sea to fish, or as air swirling the bird. "Truly, I say that Man is the Son of the Earthly Mother, and she received the Son of Man through his body, just as the newborn body is born from his mother's womb. Truly, I say that you are one the Earthly Mother; she is in you v you in it from her you were born in it and you live it again you will return Keep therefore his mild, because no one can live long and be happy but he who honors his Earthly Mother and satisfied. his laws because your breath is your breath, your blood, your blood, your bones, your bones. Your flesh, his flesh; your bowels her bowels; your eyes and your ears are their eyes and ears "truly I tell you, yes. you left to meets one of these laws, if you damage only one member of your entire body, you irretrievably he lost in his painful disease and would mourn and gnashing of teeth. I tell you, unless you follow the laws of your Mother, not able in any way to escape death. And who embraces the laws of his Mother, he will embrace his mother too. She will heal all its wounds and he never sick. She will give long life and protect you from all evil; Fire, water, bite of poisonous snakes. Well, your mother gave birth to you, keeps life in you. She has given His body, and none but she heals you. Happy is he who loves his Mother and lies quietly in her lap. For your Mother loves you, even when you give back. And how much more will love you if you are back again to her? Truly I say very great is His love, larger than the largest of the mountains and deeper than the deepest seas. And those who love their Mother, she never leaves them. As the hen protects her chickens, as the lioness her cubs, as the mother to her newborn, and the Earthly Mother protect the Son of Man from all danger and harm. "For verily I say unto you evil and myriad dangers await the Sons of Men.

Beelzebub, the prince of all devils, the source of all evil lurks in the body of all the Sons of Men. He is death, the lord of every plague and putting a nice dress, tempts and entices the Sons of Men. It promises wealth and power, and splendid palaces, and ornaments of gold and silver, and many servants. He promises glory and renown, sensuality and fornication, drunkenness and binge eating, wild life, laziness and leisure. and tempts each according to that for which more his heart leans. and the day when the Sons of Men have already become slaves of all these vanities and abominations then he, in return for this, they snatch all those things which the Earthly Mother tan 6 abundantly given. Les snatches your breathing, your blood, your bones, your flesh, your bowels, your eyes and your ears. and breathing the Son of Man becomes short and stifled, laborious and smelly like unclean beasts. And his blood becomes thick and fetid, like the water of the marshes; coagulates and black as the night of death. And your bones become hard and knotty; are rolled inside and outside are crumbling, like a stone falling on a rock. And his flesh becomes greasy and watery; it corrupts and rots with scabs and boils that are an abomination. And his bowels are filled with detestable filth oozing currents in putrefaction, and inhabit them many abominable worms. And her eyes clouding, until the dark night wraps; and they cover their ears, like the silence of the grave. And finally, the Son of Man lose life. Because he did not keep the laws of his Mother, but added sin to another. Therefore, we are caught up all the gifts of the Earthly Mother: breath, blood, bones, flesh, intestines, eyes and ears and, finally, the life that crowned his body Earthly Mother. "But if the erring Son of Man repent of their sins and repaired, and back again to his Earthly Mother; and if it complies with the laws of your Earthly Mother and freed from the clutches of Satan resisting his temptations, then Terrene Mother received back to his sinner with love Son and sends his angels to serve him. truly I tell you, when the Son of Man resists the Satan that dwells in him and does his will, at the same time are there Mother angels to serve you with all their power and free him entirely on the power of Satan. "for no man, can serve two masters. Because either it serves Beelzebub and his demons or serves our Earthly Mother and his angels. O serves to death or serve life. Truly I say how happy are those who obey the laws of life and not roam the roads of death. "And everyone around heard him his words with amazement, for his word had power and was teaching very different way to that of. the priests and scribes and though the sun had gone down, not went home sat around Jesus and asked, "Master what are these laws of life? Stay with us and teach us a while. Love us hear your teaching so that we can heal and become righteous "And Jesus himself sat in their midst and said:" ". And the other replied," Truly I tell you, no one can be happy, except one who obeys the Law We fulfill all the laws of Moses, our lawgiver, as they are written in the scriptures". And Jesus answered: "Seek not the law in your scriptures, for the law is life, whereas writing is dead Truly I tell you, Moses

7 did not receive God's laws in writing, but through the word. alive. the law is the living Word of the living God, given to the living for the living men prophets. wherever there is life is written the law. you can find it in the grass, in the tree, in the river, in the mountains, the birds of the sky, the fish of the sea, but seek it chiefly in yourselves for truly, I say that all living things are nearer to God than writing that is devoid of life God made life and all. living things in such a way that they taught the man, through the ever-living word, the laws of the true God. God did not write the laws in the pages of books, but in your heart and in your spirit. they are in your breathing, in your blood, your bones, in your flesh, in your gut, in your eyes, in your ears and in every little part of your body. They are present in the air, in water, on land, in plants, in the sunbeams, in the depths and heights. All I speak to you to understand the language and the will of the living God. But you close your eyes you cannot see, and cover your ears can not hear. Truly, I say that writing is the work of man, but life and all its hosts are the work of our God. Why do you not listen to the words of God which are written in His works? And why do you study the dead scriptures which are the work of man's hands? "" How can we read the laws of God somewhere, if not in Scripture? Where are, they written? Read that for us. Tell us the laws of which you speak, that hearing them be healed and justified "Jesus said," You do not understand the words of life, because you are in death. Darkness darkens your eyes and your ears are clogged by deafness. For I tell you that no advantage whatsoever you to study the dead scriptures if by your deeds, you deny who has given you. Verily I say God and his laws are not in what you do. They are not found in gluttony or drunkenness, not in riotous living, or lust, or the

pursuit of wealth, much less in hate your enemies. For all these things, they are far from the true God and his angels. All these things come from the kingdom of darkness and lord of all evil. And all these things you carry in yourselves; and therefore, the word and power of God does not come in you, because in your body and in your spirit, inhabit all kinds of evils and abominations. If you wish that the word and the power of the living God penetrate you, do not defile your body and your spirit; because the body is the temple of the spirit, and the spirit is the temple of God. Purify, therefore, the temple, that the Lord of the temple may dwell therein and occupy a place worthy of him.

For me it is a universal legacy of Egyptian discoveries and advances in neighboring territories that bring us into the modern era with so many mysteries to clarify.

We must distance ourselves from the current religions that condemn everything they cannot explain, they reject so complex and cannot be done internally, as a barrier to vested interests. Egyptian era Aton, not wrong to worship the Sun as a God of creation, because there they developed a system of introspection and bequeathed to humanity many mysterious knowledges. Tell the Amarna-monotheistic religions center founded by Akhenaten, proclaimer of sun worship that prevailed over the ancient religions, proclaiming monotheism on ancient traditions, to assign multiplicity of gods their concepts of worship. Per the records of ancient civilizations, it is still present as a center for tourist expeditions. Here the first system developed with a concept of divine creation and life after death dedicated to the sun god, Aton as sole creator deity manifested. A concept which does not resemble the concepts adopted by other religions focused on human gods, Christianity, Buddhism, Islam and other religions known. The Egyptian concept recognized the divinity and life after death and a single creator deity, is not the same under which other religions have been founded. The only mature a unique concept with a firm foundation in monotheism are the Hebrews from which the nation of Israel and the Jews who proclaimed arises which Jesus was heir to David, as the true king of the Jews as their God. The concept of god for the early Hebrews declares that God is the creator highest that the salvation of the human being is unique and personal, that God is so zealous in his relationship with the man who allows another man intercede with him and that salvation is solely and directly with God. Mathematics, geometry, physical laws, metaphysics alchemy and all the advances that have accumulated in these times of high and low Egypt with Greece was guided by many of the first scientists who started the theories that today we study in schools Democritus, Archimedes, Thales, Plato, a long list of scholars of all branches of knowledge. The fact that his religious ideals like the Hebrews were founded on a superior deity and not in humans, apart from current concepts, with so many developments worldwide have been enthroned in fractionated races and god's concepts that divide humans in the spirit gods and sects with different names. If that is the privilege of humans we must look with suspicion, seek a more universal understanding of the Almighty Creator. Those who tried to outdo the gods of Olympus and the fables of the Roman gods not forward anything to humanity. At least these gods were temporary and present in this physical plane. By providing spiritual advancement beings he leads to the worship of gods ethereal with the name of a mortal, because it could not set up a human idea of a disembodied deity worship as always has manifested the god of all creation. Only globally there are people who advocate a universal system of values and a common universal good. There are few nations that are committed to this vision. I hope not surprised with the predictions seen as a maximum, future event, they all think they do not fail.

121-Nostradamus: ISIS Predictions

Teurgy magic and metaphysics

My personal version in year 2012:

For the future, the situation that Nostradamus predicts for what is to happen from this writing played by me in 2012. I'm surprised that this has done to date 2016 -2017

Other prophecies are fulfilled as soon as 2017. Whoever takes the power of the American nation should ensure a plot against a new elected president of the United States or an ally who turned heads with promises to the nation, this must ensure that their history is not disrupted again. The facts are manifest in the continent of Alaska where the next conspiracy is located. The risk can be avoided by taking wise decisions. Gaspar Pagan 2010, 2012 to 2016-17

Islamic Predictions

From the bowels of the homeland of Zoroaster-) one of the greatest threat looming over humanity; hunger, the killing of civilians, the global destabilization is at the gates of these generations starting century 2012. A replica of The Jewish War Josephus A closed to the public right government. People starving in the streets and sidewalks outside the fortresses, fell like flies on top of each other, like zombies aimlessly, eyes wide hunger and ulcers showed off her bowels, while inside the fortress others show wheat and food to hungry people.

Nostradamus: 2000-2025 - By: Jean Charles de Font Brunes, Centuries lll- Page 87 -3 "Mars and Mercury and money, united together Towards noon extreme drought: In the background of Asia earth tremble tell Corinth, Ephesus then perplexed. If the tests of the different dates that brings the historical relations in the writings of Robert Ambelain, add up the prophecies of Nostradamus with events that can be consolidated now of 2012 onwards. Robert Ambelain "Jesus the deadly secret of the Templars and other works of authorship reports that the birth of Jesus might be 16 0 17 years earlier alluded to in the story of the promulgated in books teaching to the churches.
The dates are set by the Gregorian calendar. who can prove after so many changes a certain date. If Nostradamus ignore these details and the changes made to the calendars, the chances that these centuries are oriented to by events. "the war (Mars, god of war, the Balkans and the Caucasus), corruption (Mercury, god of thieves) and the power of money reign together; there will be around noon a great drought; (Money=Japan live major earthquakes, Greece and Turkey have problems, Iran Turkey and other Islamic peoples are at the gates of proclaiming Islam, if Turkey sees an opportunity to prevail
Islamism could go against the West. (USA)

Prediction II, 46

After much larger human meeting is prepared, the great engine ever renews; Rain, blood, milk, famine, iron and pestilence. Fire seen in the sky, running big spark. After a large gathering of troops, a larger one is prepared after the revolution (rain, blood, will put an end to the good life. The iron, war and epidemic, then a large comet or body of fire crossed the sky. Burning torch in the sky will be seen, near the end and beginning of the Rhone Famine, sword, late relief provided Persia again will invade Macedonia. Macedonia is a military target Iran conflicts of the European union, Turkey will see in this an opportunity to attacking Greece for its conflicts of years and yet be the perfect excuse to give back to the West. on the other hand, the prophecy of Orion is based on the same data, the official calendar established by religions and accepted worldwide as fixed. the final will be 2015 to 16, per lunar of birth calculations Robert Ambelain and date of Jesus death.

It should be noted that Irenaeus of Lion puts Jesus dying at age 51. This finding someone who was present at the time shows that it is possible that manifests. This is a logical deduction, time adjust else.

lll, - When next the defect lunar. One and other non-distant greatly, cold, drought, danger in the borders, even where the oracle has begun the word, lunar refers to Islam because of the moon. (Half-moon) which is its symbol even in France where the oracle has begun (country of Nostradamus) when Muslims are about to commit a fault (of conception between them and the West) too big, cold and drought will be known, even in France I, 67. Note: There are consecutive quatrains that make a chain of events that intertwine and point to Islam as the protagonist of these prophecies the consequences of an attack on the West by Iran and Islamic countries and potential allies that. see an opportunity to strike at Israel, Judaism and Christianity included, which incidentally destabilize the American nation, being its allies. In this 2012 Iranian President, is focused on those strategies. The governments of countries where hunger is an epidemic for its overpopulation, see in this event an opportunity to resolve the problem of overpopulation, and as burnt offerings history of repeated human beings would be the solution to their problems. That is another nation that, to fend off his attackers, causing the debacle and extermination, a legal way to exterminate the HUMAN surplus that cannot be fed in the future.

A sensible solution to this global situation would be a cooperation agreement where oil producing nations, freed prices worldwide for producing larger foods, cheapen production costs and can double global food production and this is supply to overpopulated nations.

Not only the United States is called to solve the problems of other powers, each country should enter agreements to release the barriers that stop the growth of food-producing countries. While being used as weapons of power, stopping the supply of goods as a means of pressure to solve a global problem. Nations will be doomed to receive the same treatment. Function as a funnel as before where solutions fail to pass as adequate slope to the problems that each country faces ... catastrophic consequences for all nations are coming, be resolved if leaders worldwide not demote their attitudes towards the planet, this is his reward. Gaspar Pagan (Edwin)

As a whip that refreshes the memory, history repeats itself and beings who have not learned the lesson must face debacle and destruction, or return to world peace. Everything created looking for its balance and green grass born the fire should consume dry and the universe is renewed in every action, which leads created its very beginning to emerge again a rebirth.

125-The mystery of acquired knowledge

Stop Earth from her womb similar shapes and different forms of creation. For example, a small wisp of gas that emerges from a chemical reaction contained for thousands of years within time out of his cabin where he remained isolated and combined with other energy sources: hydrogen, phosphorus, potassium, arsenic, sodium, etc. Of these combinations and the internal heat of the Earth stones, metals and other combinations which, when mixed with elements of cosmic emanations create a variety of new forms of expression arise. An agate stone is just that: a portion of gas trapped inside of matter is cooled and continues to grow capacity. It is an example of the reactions of creation that cools this continues to grow, locked in a niche that created the earth itself. It is part of the reactions that are repeated and taken as laws to explain phenomena that occur.

Death, change, transition. When these powers cease? What is this mass of energy that went and where it goes? What law regulates the period between a phenomenon and another? - The reorganization of matter in so many varieties of expression that surprises us every day more diversity in which it manifests itself. The mysteries that encloses the evolution of species over millions of years kept busy scientists, thinkers, mystics, archaeologists, paleontologists and all that branch of knowledge that seeks to unravel the evolution of species, the human knowledge largest chain, which is the bearer of its own information and was unable to decrypt its contents so far.

Our inner being, the soul and conscience of it, when the soul rises to understanding where he was kidding and takes note of the attributes that generated the mind that created it assimilates its shape and gradation. That is the legacy that gives freedom. The pure energy of the creative vibration degenerates to act on the scale of the demonstration that a greater or lesser degree contains. It is the way to make themselves known, to act on the descending scale and take that degree of existence, to interact with other degrees of energies that were created by the very emanation of what was the first source that emanated, which is manifested. Light cannot exist without darkness. Life cannot exist without death. All forces have in its emptiness, its opposite. In the mind of man are all dimensions of creation and is the only one who realizes and performs. If we are an emanation of a creative principle that encompasses everything, it cannot become aware of itself, unless it has a mirror that reflected and self-known. That is the reflection of the force that must emerge and realize that it cannot shirk reflected in its opposite, or lose the quality of manifest and affect a conscience, an entity gradation whatever, that make or to make it available. Ego strength of creation, his passion to create would stop or be distorted in their duties. Purity cannot be a world of existence because it would nullify itself within its own energy and would only be a reality for itself. The essence that contains and produces would turn in his own cabin and contain herself. No need to be known, would not have to emanate or share your essence with nothing and, therefore, their existence would be herself and nothing would perform. The laws of the universe are changing and rest, everything is in constant motion. The allude to forces purity is back to the center from which everything emanated screened or inadvertently disclosed. The fact that you want to catalog or give a description of generations created by the emanations of this primary force should crystallize rather than evolution must return to the source that generated in its pristine nature, as emerged from the original source. This would be the biggest dilemma that being must face. There is no natural or spiritual law containing that notion into being. We can use all sentient powers to wish that stage regression to primary evolution and would not be possible, nobody would make it. The energy itself that created us lost control of what it generated. The make themselves known is the ego of creating itself, it is a necessity or impulse that leads to follow that generation of effects and manifestations in the universe, being is your vehicle. He stops in the middle of all creation, we realize, we look back to that source, which itself is canceled by opposition, the midpoint, rest, neutrality, where all establishes a balance. The opposing laws are attracted to create, this happens all the cosmic universe that created us, the channels that separate all forces to maintain a balance. The same between planets and atoms and all materials that meets the universe that we cannot capture in our finite minds. The kingdoms created are like the flower that languishes and ceases to distribute its perfumes, aromas and colors, then it becomes his seed; it contains it and returns to its beginning, again usually another plant with the attributes of the above, but may not be all floors. This must be the wisdom of being when it surpasses itself and becomes part of it.

Emanating energy cooperation towards primary source for an internal maturation of knowledge the laws themselves is a channel that gives us strength to achieve harmonize our inner world. It is not pushing but we managed to attract harmony. It is the rejection what gives us the victory, but the acceptance, the attraction of opposites; that is the overcoming of chaos. When unilateral forces are weapons of power, are a bottomless pit, wrap all creation and destruction knows no bounds. There is no way to feed it consumes all the energies of the earth and has no boundaries. They are the powerful unscrupulous who have managed to entire civilizations and lead to their own destruction by the love of power. The love of power only reveals the weaknesses of sick minds, who do not have the courage to recognize themselves. They need the power and domination for others to suffer their weaknesses. If they gain control engulf humanity, for its purposes prevail. The positive forces of creation cannot be out in any way, no power to stop them. Greed is a bottomless pit; the sufferer is born with a worm inside and be consumed by it, like the notions of sin and guilt. Acceptance of inner demons that flood our nature is a wrong that does not allow to be raising their spirituality and overcoming concept. Accepting that we are in developmental stages and each incarnation is an opportunity to overcome this condition and baseness to which sometimes succumbs being is the most sensible guide to redeem our inner maturity and raise our soul to more purity that is the right path. With all the opposing forces when an intermediate gradation is achieved peace is achieved. The high, the low; sour, sweet; peace and war; wide and narrow; the right and left; life with death; Light and darkness; movement and repose; love and hate; exterior and interior. Only even-handed leaders committed to strong values to defend a position of global harmony, with the ability to submit barbarism, that is the purpose of existence of beings. We are the sephirot of human cabala concentrated in manifestation, the Vedic fields that can be studied at different spiritual philosophies of life contain elements that help to mature sublime concepts of our powers of healing and divinity. From the ancient Hindus and the heirs of his teachings, Zoroaster and many of the old traditions that contribute to the understanding of spiritual life to strengthen a comprehensive concept that is not focused in one direction, for a universal vision of divine energy supporting the creation of all genres to which we have access. To achieve the strength of any emotion or factor to guide us in that direction, you need the union of beings who communicate with the same principles and purposes of overcoming the meanness to which we expose the weakest. We cannot abandon the passions, disharmony, we created us as beings. Tolerance makes the most even-handed laws. "But beware of violating the precepts of universal harmony, its axis could change course and all succumb, and human power no longer would need". On the contrary, nature is a world of attractions that never satisfied. Everything that comes within reach will use it to produce something different, his passion is creation. The primary substance benefits everyone, no one has. If someone would possess herself and could not exist in the same plane, as it would destroy itself for being opposites in a fourth dimension, a black hole. If it becomes possessed, the ego of the possessed not share with anyone, would build their own castles and alienate of other mortals, as a beautiful woman. Just feel that it is his master would his slave. It organizes all that exists, makes operating laws. If it spreads, re-organize, create new forms of old, used those same atoms, electrons, neutrons, photons and every particle that crosses their path to produce, create; harmonize the whole universe, His law is love for creation. You enjoy the beauty of form, the subtlety of what impresses our senses and the senses of all creation, perfumes, color gradation that impact our admiration. If we imagine a new way, he insists that is created; his passion is to please and impress the perfection of human beings. All the beauty of forms is organized by the need to express what you want; what attracts our senses and admiration, it is organized per the needs of harmony in creation. In addition to the other realms that use their own laws of attraction that its evolution is successful. Love the desire to create perfection, the union of opposites; establishes peace in the universe, organizes its future with laws that must duplicate or change its course to follow progressively, harmonious with its emanations ... Being cooperating with these laws is peace, inner harmony. It can be seen in the divine plan a passion to exist, like other creatures who praise creating their songs, perfumes and colors. They dress in colorful plumage to impress, attract their peers; flowers produce the best perfumes, elixirs that please Bee and insects to cooperate in his plan to reproduce, or create more of their species. Fruits that contain tree produces the best dishes to our liking, love animals to consume them and let your seed to reproduce elsewhere. Endows the nature of fluids and ways to excite the passion and desire of opposites. This need arises overcome his passion for creation, use these intelligent measures because they do not can move and the same need gives them these skills. Since originated, he organizes the universe with its main laws; It produces and creates new, to fulfill their creative mission. What nature destroys it reconstructs and repairs; It generates new forms of nature destroyed. After the fire burning new life arises and exceeds the previous one. In our body, harmonizes violations of our appetites and repairs damage we create with our violation of natural laws; clothes with our imagination the total of our expression, although we do not realize. Penetrates every corner, he directs their creative energies to establish the harmony of the disease, which is created with the disharmony of its misuse. When violations of its laws are extreme, heavy penalties, destroy everything that has been created, violating the subtle laws for freedom of choice. We wonder what we perceive as beings, we complain about what happens around us, that the laws themselves punish violators. It gave us the freedom to choose and violate the most elementary laws. If we are aware of that truth, we must seek internal improvement and come into harmony with its laws. The cosmos they apply the same laws of reorganization. Are our allied laws that create harmony and perfection? But they can be the most destructive forces if haywire. The wind with its slight breeze brings us the elixir of life, share its essence with all creation; the same as the fire, which gives us warmth and cooperates with the way we live and exist. If harmony is lost, it happens as a destructive thought devastates and destroys everything in its path. The power of thought affects our system and our environment; so much so that the great masters tell us we are what we think. The laws of creation are harmonious with the energies of thought. These attract or away peace and love within us; You can create or destroy our own inner harmony. Opposites are part of the heritage of the imagination, we must understand the laws that exist for the best. We must know and choose what is beneficial. One who chooses evil, negative, suffers the consequences. Creation is only friendly with everything harmonizes with it, but does not distinguish when to destroy. His passion for creating never fills completely, all puts a limit to exist. Everything succumbs to make way for the new creation. All waver at the beauty, all aspire to possess, but the ego consumes them when they become the envy of others. If you have, watch day and night not to come thieves steal your treasure. It is the balance that puts your judgment on what dish put your counterweight. If you forget, you will ruin the imbalance. Learn to master your passions and focus your attention in the middle of the scale; weighs your actions, measures and tolerates else, I am compassionate and friendly. Everything pursue the ultimate in our life, we seek to be better, happiness, the purest love, exaltation of our being. We attribute to achieve. Most look inside and manages to rip into the walls of time something that is perceived as superior. But not having a notion of what is sought fall into a crossroads of concepts that tie their freedom of thought, their free choice. Whoever discovers the meaning of these inner realities should not give them to any mortal. It is a treasure that belongs. The day I delivered, only part of the happiness of another and depend on it. At the time to stop thinking or seek their freedom, only you lose their dependence on that source. Be free, be yourself. Even the universal laws operate on that principle. Others will embark on foreign search in the material, in amassing properties and then discarded because things do not fill them; He dominates his ego and are victims of their possessions. When you reach the maturity of your consciousness you do herself, you give light with the light of the universe itself and you get one with him and be you own light, you will be eternal and matter no longer be necessary. Within a man of light there is always light, and the world will be reflected in it.

The Egyptians

The Egyptians were connoisseurs of the secret of creation lay where Akhenaten superior wisdom to his race. The day the Egyptians hidden secrets are revealed the world will realize that developed the knowledge they possessed in ancient times and awarded them the Egyptians a role worldwide recognition. Pharaoh Akhenaten came to the melt material being with his knowledge and clear insight to perceive the divine causes the human being and the highest. That same foundation cone was acquired by many disciples who were in contact in the ancient schools. The legacy of this pharaoh who introduced the concept of monotheism in its time produced an abundant teaching school that has been treated to wipe out the Earth by those who adopted their teachings. Even his priests rebelled against rob them of divine authority they possessed. But despair and destruction they turned off the light to the next world. The Corpus Hermeticun, sun worship based on the Egyptian mysteries, says Massillo Ficino, he met a new splendor. The sun embodies the divine light, spiritual enlightenment and body heat, in descending order, to God. "We are between two furnaces fumes and emanations of God summons the whole universe." As above, so below. We have created measurement tools for multiple forms of energy traveling through space; make a spectrum of all elements moving back to Earth. Men of all branches of knowledge, observers and mystics seek ways to give mankind knowledge of the laws that affect it. This creation is led forces we perceive with our five material senses. They have demonstrated the wonders of these measurement systems. Musical scales give us another notion of these vibrations we can create in our physical plane, which can reproduce knowledge and personal domain. Vibrations duplicate to use as a harmonious manifestation of our perception of the cosmic scales at which we have access: light, energy, solar emanations and, beyond that, the projections of cosmic emanations that reach our planet. Scientists have overlooked a simple law of nature, which is the creator of all these manifestations, and the need, the passion of creation itself. Throughout the universe, if something comes up and says it is because there is an emergency of some kind in the universe itself to regroup matter what manifests itself physically. We have reached the domain and knowledge of the infinite matter what division: the atom with all its components. The cohesion of all its parts, by laws of attraction. The separation, by laws of repulsion. And a third law unites, characterized by the harmony of neutrality exist, emptiness, the midpoint of meeting any energy cosmic harmony that manifests a product of that interaction. For these energies, manifest, must exist in the same vibration of matter, energy superior to all of them that gives them presence in the universe. Being negative and positive, as we have listed, they must obey a third force that harmonizes force or to manifest and create. Imagination regroups these energies and shapes them before they materialize on the physical plane. That is a force that exists only in the divine mind and being and the nature of universal cosmic level. An inner drive that must be plagued with energy pulses of quality and harmonizing it within our subconscious mind and then stop at the conscious phase, where the urge to develop that creation fills us with immense satisfaction. So, great is driving a global economy at all levels imaginable humans. We should remember that when we speak of being in the sense of expression must be understood father-mother God, men and women acting in unison, and the Great

Father and the Great Mother emerge as attributes of God himself. They are elements of the perception of the ancient cultures that were closer to the true manifestation of creation, which has been lost in the thickets of the interior space.

It is a natural intelligence that directs all that exists and acts from its most simple or complicated form of expression, that being part of what we are we can perceive even part of us and the universe, because we it and we project in energy to the source that gives us the characteristics that physically occur in the material plane. We use a subtle force of attraction, a magnet or static force so that energy is present in this plane and manifest for periods of time. There should be a gap of this somewhere in the universe, with all the attributes, which achieves by attraction descend to the material and spiritual reaction, which then exist for periods of manifestation of creation in this plane should return to its original shape. The energies of matter are incorporated into the material realm and the soul, the sublime matter of the soul of creation.

It has been since the beginning of time knowledge that only a few have gained access. A secret of thousands of ways to describe veiled by those who managed to imbue that link the soul with the consciousness of the Creator Father. It is known with certainty that it is a place where the soul ascends and climbs through a tunnel of light and turns back down in a dark tunnel. While traveling back darker it becomes. This experience has many human beings who have gone through a process of personal resurrection. I personally know a humble person who went through that experience and thus know God how many thousands of people have had the same experiences that fail to explain and stored in the depths of his soul, because incomprehension scares them and do not dare to testify.

136-The maturity of the soul

The light:

Analyze a matter like this would ask where is the light when turned off. A power switch is the mechanism separating a reaction of the other. We only your temporary manifestation stops flowing, but the power remains latent until it actuates the switch again. No matter how many times it done, energy will manifest. It will be shown the capacity for which it has been regulated. We have an example in the bulb compared to this phenomenon. Once the service life of the bulb arrives, its material should be replaced by another mass. In that span of time it takes to change, energy remains latent until a new body takes the presence of the above. In creating is an accumulation of infinite energies that can take as an example the electricity and has the capacity that it can manifest many creatures without limiting its ability to manifest. The same in the electrical capacity on a single source can be connected thousands of bulbs. These are laws that man can control and channel their knowledge, but cannot create or destroy energy; only to unmask and will be useful. It is a clear comparison to be have a better idea of what is the creation of beings. The statements of spiritual laws are derived from my personal experience to be mystical knowledge acquired as a student of law ("AMORC") Ancient and Mystical Order of the Rose and Cross.

The only institution that exists worldwide as a school of initiation, to disseminate personal knowledge, on such profound topics dating back about 3500 years. When human beings of creation, which can express the Father's kingdom inside emerges, it is known that his destiny is to project the divine spiritual heritage that will accompany him during his period of existence in this earthly plane, that the divine plan god who created us. Once detached from the material body will join again and diluted in a scroll regenerate energy until it is drawn back to this earthly plane.

The mystic who knows the laws of creation knows that once the soul leaves the body, the darkness, which are always present, take the place of inner light and matter is projected into another dimension, like vibrations what were the attributes of the soul. The mystery of the integration of the soul or subtle energies that emerge from the body does not lose its qualities nor fade as is the common belief. There is a containment level or dimension of these energies into the soul-personality has access. sky, cosmos or the name assigned to the diversion of natural processes is called.

Beings who are preparing in that regard can consciously enter this area or dimension of divine creation and abide by regeneration periods to return to this plane of expression as a new living being, a new child with a new soul, a new Living being.

Reincarnation, that is the legacy of Father Creator and the teachings of Master Jesus announced in the Gospels of Thomas and Philip. This act is a subtler vibratory rate that raises the vibration of the body mass arises; the laws of attraction and repulsion and magnetism in being arise. This is what keeps the inner universe in constant motion, while exuding new manifestations that occupy the spaces and create their environment existence. If someone imagines the other forms that manifest themselves in different species, each vibrating at its ability to attract the primary source of energy that complement their form of expression. Only human beings can perform a higher world where the divine laws of God harmonize with his soul-consciousness. The attraction and repulsion are two qualities that act and a third force is manifested in the harmony between these two opposing forces and emerges what is called love. To understand, it is a law that governs the energies where a neutrality that harmonizes the forces is maintained. Harmony or balance of these is what we call love or union with each other, without mixing completely. The force that brings together all the forces within a small universe and in the grand universe. Above all, this a force that vibrated as an emanation arising as a scroll or effluvia or emanation- forces interacting with previous manifested cosmic sphere. It is where, by the interaction of all these basic laws of creation, a unique and true dimension of being arises. To mention being is the broadest of creation paradox, the emanation of God.

Being A three-dimensional scroll vibrations groups, involving energies of primitive matter in being. Upon contact with empty subject of its contents, they perform a combination that gives them a definite strength (cohesion) and sets in motion other materials. From this combination, a dependency arises. The core that is created will still exist while having access to

these cosmic energies. It's like a fourth dimension, where empathy creates the field where being cannot penetrate; only his talent emerges and share without mixing radiation intelligence so that it can interact with them without being able to draw it fully into our dimension known manifestation. An indescribable emotion that seems to cover everything we projected, and we cannot penetrate, but it shows us and we perceive their existence and as Archimedes said geometrizes God.

The brain components of atoms are coming into this small universe of physical reactions. Of these mergers misting of subtle matter that gives the material properties to all that exists arises. What we call conscious mind in being, it is the attribute of that energy. It gives us the ability to realize the movement, time and space dimensions, an intuitive knowledge of its content, because we herself. Arises somewhere dimension of creation, either within us or outside us where these combinations are drawn after creation. Comes a cosmic dictionary our relationships and emotions. We assume control to express emotions and describe them as laws of principle. Likewise, this reaction and combination of energies is maintained when being stops vibrating in the material body. These are restored to the cosmic part of which arose, and rich in emotional maturity is part of the universal soul or substance, files (Akashi). Internal understanding of that world adjacent to the physical expression, is the cluster of mansions soul spirit energies that are added energies are purified by new experiences and maturation of the benefits of what is pure and divine. The repetition of these states to overcome is what guides us eternal life, because these energies are continually manifested in new creatures and returns to the cosmic consciousness of God is. Only the word we use in our spoken form gives a sense of understanding and action in our minds that envelops its content. In this vacuum conditions are created in our body to produce the appropriate time to give a sense of expression to what we imagine sounds. Those who first discovered this communicate them to others in spoken form. It must have taken thousands of years to form a vocabulary that were cataloged to communicate. The myth of the cave is an ancient example that describes how the imagination of a human being is unique; It has a wealth that can share, recreate and project the same images our other minds that are unaware of the phenomena.

Lighting

Being that achieves harmonize with its light on this earthly plane will have the blessing of intuit and will not need to see it, because it will be part of it. If your light reaches our inner and projects in our consciousness, being projected by a space-time tunnel, where it reaches the site where the light emanates from the presence, the portal of light. The soul, which is the product of this interaction where the subtlest emotions and comprehensive primary cause us belongs staying temporarily and let us enjoy your surroundings and realize its depth of happiness to which we are entitled to aspire.

Today, with technological advances, the story has projected us through the centuries constantly changing as we dealt with new discoveries. They discovered evidence that change every time some of the realities demonstrated so far. The story must update so much data that have been taken as certain that should be examined very carefully the text and start again. This has happened before and repeated cycles. Intellectually dedicated so much energy to create a world to save for posterity the experiences and results that are floating in the air and what we know often crumbles before our eyes. We realize that what is taught is plain content, which is imposed meaningless knowledge that teachers teach what they learned or what they are given as a matter of education. Keeping our inner world killed by confusion created by concepts of repression sometimes pushes us to see ghosts in worldly things that have no influence on our spiritual realities. Being that is influenced by these teachings that degrade its condition of being Despite centuries of social dislocation, have emerged figures worldwide who have exalted their convictions in their ideals and have led crowds to have a belief in something that feed them to follow the search for truth. They have fought on the side of the weak and the poor, they have laid chair of devotion and humility. They are enlightened who have spoken in favor of the ideals that have accepted as their north and have fulfilled their mission. They are admirable beings and have transcended history as wise and humble in his convictions. Less educated in school, but

with a commitment to truth and the struggle for human excellence must not waver. You must realize that global humanitarian leaders have given the daily struggle for humanity is exceeded.

Intuition

The statements of the creation of self and all that exists in the universe will never be captured in its fullness by the finite mind of man. We can intuit a vague notion of what could have happened at that time. No history book that educates us about these issues; with the only thing, we have is with writings that tell us the stories of the characters who were involved in the events that were documented to give an example of life in those times. Analyzing this aspect of human knowledge in the field of metaphysics is as close to what has come. The stages of knowledge we have built and accumulated a story being met personally demonstrating the rarities during those early periods were being projected some extinct and unknown qualities for those who inhabit today. The most profound minds have dedicated mental and physical efforts to even come close to a rational stage of something that seems logical. For more descriptions that can be written, it is knowledge that will conform only to the mind and the conscience of whom describe it. Great thinkers have given us knowledge that we must update and continue the search with these bases. Looking in the archives of my memory I find that these were attributes of mind needed no preparation. Intuition attributes are based on the mature aspects be at times where only

dangled from above to capture any phenomenon to happen. It was a magic to many who were not trained to live in places far from those populations. When were, these reactions were surprised to see, that things were about to happen they said. Hence arises Magic, divination, for those who saw how people marveled these powers. Seeing that it was a condition that could be harnessed to impress and charge for these tricks, it became fashionable to prove the charge phenomena that the mind could not explain. Today we see magicians and troupes creating tricks to please the minds of those who love magic. True intuition does not lend itself to such demonstrations. This was an attribute of precognition of people away from civilization and he used this ability to orient themselves in their daily lives without realizing they were doing amazing things, for them it was a routine.

142- The original lost knowledge

Much has been lost by manipulation and interest created to measure. When the first consciousness was manifested and interact in what we associate with a Creator God and all the qualities that every being overt or perceived, and ascribes to a scale of understanding their functions per race and the values that practice. They agree to the possible imposition of a leader at some stage of domination and similar items of that culture. From the most primitive species to be developed all that is manifest in the future will attributes and properties of the same universal mind, its laws are the basis for understanding life on Earth. He

legacy of beings who devoted themselves to documenting the ancient history is another flow for us. From the time of the Egyptians, the sundial, the construction of the pyramids, equinoxes, mathematical statements, heliocentric theory of Aristarchus 230 a.n.e, and all the knowledge gathered in the great schools of Alexandria in ancient times. With the destruction of the Library of Alexandria, where they were cataloged the most extensive knowledge of the minds most advanced of humanity, scientific, historical, mathematical, philosophical and all branches of knowledge discovery (many) have been lost. We do not know if they were stolen and hidden, or truly burned. When it was discovered this knowledge by the Christian Church and saw the growing threat hanging over her, she hastened to send a mob of fanatics to raze everything to realize the mistakes that put into practice, trying to make the deification of Jesus. sun god incarnated in person, by Saul of Tarsus who believe a god to the Romans and their intention to take over the empire, of Nero. Who has doubts to seek and find. The important part is that mental disorientation and reciprocal approach to energy beings react to the arguments focused on an erratic culture, misaligned forces and directs them to concepts that form a bubble fanciful creations. From that source continues to fuel the imagination of thinking beings, minds began to be derailed by distorted paths. What the mind becomes realities of divine energy and its variations coagulates in a reverberation, this will continue to grow at such a level that the reality is updated every day will be absorbed by most beings to return to a new stage of maturation. Children come into a troubled society for their erratic concepts, which have matured that civilization will disappear. The legacies will manifest themselves in the loss of civil orientation. The repetition of concepts unclarified full pro leads the imagination to create internal forces that eventually lead to uncertainty. He has awakened from the torpor of trends manipulations created. That should be the awakening for future civilization. Genetic manipulation, subtle cosmic energy, energy emanations from outer space manipulated by those created for the welfare of human systems, destabilized the forces that dominate and control the balance of creation. We are treading on the margin to control and manipulate energy that has made succumbing often the nature of the planet. The imbalance of forces would lead us to what would be something like Atlantis and Lemuria. The common man has not matured a way to harmonize with the energies comprise balance your thinking being with variations of natural laws. Restlessness, fear, failure to obtain a security entity who promise to protect the manipulation of information by the chains of transmission that have no control or prohibition from making statements sometimes baseless, could result in an unprecedented cataclysm. If I mention here of these social elements, also the fear of death is yet another incentive for those attitudes. Therefore, that these statements are in this book is to alert citizens if there is any change worldwide in living conditions, will not be the end of the planet, much less to humans. Which make a statement like that is untruthful? No one has that power. Only the creator eternal enables us to discern how we should act and not be afraid of threats. By Patrick Gerri statements, author of "The Orion Prophecy" these phenomena have happened before and we're still here. His statements are accurate and credible until time proves otherwise.

The inherent qualities of each separate manifestation are transformed to the needs of development and the pitfalls that must be overcome. In the case of a lizard, to name one example, its development requires train engine means to overcome obstacles, a tail to maintain a balance of movement, properties in their legs to climb on trees and countless features that have evolved according their heritage and survival needs. All beings respond to the same needs and are adaptive evolution that turns on its needs and develop the region. The plants keep a genetic code of all the attributes of manifestation and duplicated in their seeds, and continues its survival. All that is of a different nature creates a disharmony in the act of creation and this is where the disease manifests itself, the disharmony between matter and the laws that must take its course. The symbols with one or two snakes observed wound in representing our spine where screwed up our brains show an expression of the two opposing parts of our body and its energies contrary gather in the center of the brain, and if the disease occurs attack each other, and hence the destruction of internal diseases. The Egyptians, Greeks and Romans ruled this knowledge and left a great scientific and medical legacy to humanity. The foundations have been created in terms of material heritage of being. Science must deepen diseases since the beginning of the evolution of being to focus their quest to restore health. The land stages of magnetic gravitational energy are disorienting; it reorients itself as animal species. For some reason in the same way to operate the human body is not guided by itself, for that reason the ancients had control of that information and practiced concentration exercises and guidance of their material bodies to restore health and inner harmony with cosmic forces. It was a virtue for them that spiritual legacy that towards beings endowed with superior knowledge. The material body evolves and creates methods to combat diseases and evolve internally stronger ways to prevent the disease. I mentioned before, it is that once any species find a way to modify their structure will evolve to a higher survival in manifestation species; even the first creatures have developed other species, derived in other different and other qualities. Of the first creatures, they have developed other species, derived in other different and other qualities. In its original environment where all forms emerged and varieties sharing the same facilities of water, and which by their nature evolved under surrounding conditions with each other, and ovulating and expelling them outside their fluids reproduction, where there was no law dominant procreation, because the DNA of evolution were organizing in the early stages of development. The fact that these materials were combined without control is the basis of diversity of creatures that populated the first manifestations of life. Climate change and storms that have affected over time disperses life in the oceans and all areas of the planet, from which emanated the qualities that have enabled us to an evolution than other creatures. The glands that emerged from the combination of sodium and potassium where the quality of attracting emerged was created and stoked the first cell that supplied protein for self-expression provided it with the capabilities to build its own means of development and evolution their primary skills to attract different conditions in the composition of the elements at its disposal. While additional cell division stoked with the energies that were gathering with how to generate new and different energies. When agitated with neurons created in this first vehicle of expression and combined with other elements that were developed in unison in the same environment and so on more complex forms. From this center as gland provided primary neurons that shared the sunlight that provided them with the necessary elements and strength of expression to generate what after a long period became man, endowed with a different heritage the other species that were created in unison with being.

The way

If human beings carry out the potential in your

inside and freedoms inherited would realize that it is holding access to the Kingdom of light and dwell in it and return it to life, and return to the kingdom in his departure and change of the substance of life and death. His soul does not disappear; only changes of expression. A state of reflection to another and final integration in purity to your destination, which is the sanctification and communion with the divine substance. He receives my word enters the source of my being, because I will get me as knowledge and put it into practice in his own being. Once you enter in communion with the nature of the Father, all will be revealed; its essence will penetrate your being. -not the spirit soul is the part that is associated with human emotions, from the time of leaving the material body and are integrated into the nature of God: love, compassion, mercy, and so on. The laws of creation take mental stage to make way for a slightly more defined and elevated processes that come into play in the maturation understanding, and define laws and principles that mark a true definition of internal processes that dominate the creation of God.

Those who do not have a comprehensive knowledge, perception of time and space and all dimensions that being attributed to what he perceives as the manifestations of what we call God. Inner stillness, the midpoint where there is no conflict of any kind, the abandonment of ego, ego, being in the middle and share the flow of all that is manifested and perceived internally. Nor is music, nor water, nor river, nor spring; It is when neither is man nor a woman: he is one. God consciousness has no sex; this is a dimension of the attributes of the human mind. To attract the divine substance of God must bow down in the center of being and surrender to

deep silence and peace. Where it leaves the I to give way to higher soul that permeates the universe, where cease struggles domain. It is integrated into that dimension where everything is one thing and the light created is confused with the material. Until that level of understanding the teachings of Jesus rise, for he was endowed with an understanding of the nature of the Father.

The aura is a field of magnetic physical energies of the nature of spirit that emerge from within and are perceived around our physical body, regardless of its structure, is material, plant or animal. Its projection is like a rainbow and covers its colors on the physical plane; They correspond to heat and melt the energies within, effect of the combination of material elements. On the spiritual they merge with the energies from outer space and form an interface between the source of emanation and the universe itself. They are components of what we created. These energies can be controlled and full of energy at will. Spiritual and soul maturity depends on the control of these inner qualities. It is the vibration of all interior that emanates to the outside and mingles with the energies of the descendants of the plethora of extremely high energies of the universe.

Compassion

The interior system of our being compressed and groups as a sneeze of all the energies emanating and are attracted in an act of protection to which we perceive as the weakest and which in turn is a victim of an act of any nature that evokes a need for protection, that suffering is shared, spiritual support is provided, depending on the depth of suffering. When the causes are spiritual and human being feels helpless for any loss that affects his soul, one's own spiritual laws into action and soon being recovered inner harmony, with a more sublime and deeper spiritual knowledge. The level of suffering prepares the soul as an initiation to broaden understanding and evoke that feeling when others suffer in front of him. Passions to be mastered

149-The Inner Master

It is one who has overcome the passions, ego, and understanding of other beings and their weaknesses and sufferings. It has surpassed the initiations of life and consciousness projects him as a teacher. It is also the one who has passed the test of having all the passions and meet human emotions. Overcoming the ego is the mission of the inner teacher. If it is stated that the thalamus gland and our brain system are those who keep these correlations and dominate these manifestations we attach ourselves as thinking beings, we should assume that mastery in humans would be the dominion and control of these areas of emotions and passions that we linked to our internal development. They are in this earthly plane many who want empower of this freedom of choice that we inherited from our Creator. We must admit that a high percentage of beings do not possess those qualities of inner maturation and depend on others to direct them. From the earliest civilizations, there have been teachers who are dedicated to guide the humblest. But they have not been so egotistical as to seize their freedom to be themselves.

Agatha Stone

It is a different world where human life does not exist. It is a simple stone that grows from inside to outside with a hard shell that does not allow guess its content and continues to grow for years and years. Follow the laws of gravity, but what distinguishes it is that, despite being influenced by the laws of attraction of gravity, expands outward like a small universe. Begets and creates new forms in its center, follow the circular lines, curves, and a range of colors that do not have a definite pattern. Demonstrating their inner qualities gives rise to everything that emanates and is discovered inside. It reveals a process that should be like what created the universe itself and a combination of gases interacting with the energies of creation, have created what we now categorize universe. Apparently not follow quantum laws that govern the universe outside, because it is its own universe. The gases were trapped inside rise to a new form of manifestation. New for those who marveled characters from antiquity to the present by the variety of its interior, per the amount of accumulated gases and formed in a bubble that determined its colors. To me it represents one of the qualities of the universe itself to be generated for the first time. The rise of gases from the first sneeze of the universe and the energy mix that flowed from these first emanations give life to a world that expands with repetition of a primary law, of clothing all layers of maturation to cool and protect your internal content, and was going from outside to inside. Once temperatures dropped, the vapors are condensed and were trapped inside and began the real creation miniature of the universe. The rest is history. Vibrate alone to their own will and ability of existence beyond the reach of many expressions, that open passed as another element of evolution. Great men have been excited about his odd behavior, their bright colors, mystical vibrations attributed to him. Connoisseurs in antiquity attributed a meaning which are engraved the oldest knowledge of civilizations. It emerged from the depths of the earth; was given birth by his own will arise, like the universe itself. The human being he discovered and used to publicize the history of mankind in its principles. They were glad to have an element which could preserve symbols to express knowledge and curiosity challenge posed by the invention of the first inhabitants of the planet. It has been handed down to us a power of the human mind in evolution, the knowledge gained, knowledge for new settlers we develop more complex systems. Those who dominated the deeper knowledge were bequeathed to us by history. They worried by exalting the most important details of events and knowledge for future generations, just as the universe itself. Contemplated phenomena in nature extol us and we are concerned today as our predecessors, and they were closer to the truth of things, demonstrating around and we inherited their knowledge in their allegories. In the construction of Solomon's temple, it is one of the stone which the builders rejected, because this is part of the material that was used to build the first temple, where temple symbols and sacred teachings were recorded, and mistakenly left abandoned. It had recorded the most important teachings that tells the story. "Show me the stone that the builders rejected; this is the cornerstone". It was highly revered by the ancients, who considered the stone of science. It applied in healing remedies in Egypt. It was used in the pupils of the eyes in the images of the gods. It was used in prints, jewelry and famous buildings such as the Taj Mahal in India. Inside the stone. In traditions, it is used to cure because it absorbs negative energy, which should be your source of growth, as happened in the first stage of the universe; when mixed duality creates the trinity and the second law of creation, from which emanates the third physical law, a manifestation of divine mind in the first sneeze of the universe. A stone that can absorb negative energy inside and combine with other energies. In this combination of energies, it is like a recorder information, like a computer, and can record intelligence as a GPS of all energies in nature and energy entering and leaving Earth. It's like an encyclopedia of energy, and if you managed to discover in it the key to interpreting this knowledge, we would obtain part of the history of the universe. The pieces that are used to store knowledge in computer chips represent a challenge for modern technology inventors. This may be another key intelligence undiscovered. This is the story of the stones to humans. It reflects the cosmic on the planet, and if the engraved stones in knowledge could be rescued, we would have a more comprehensive knowledge of mankind. The history of ancient peoples surprises the modern man, many of his stories, which occur over long periods of experimentation of phenomena that are incomprehensible to our scientific advances.

The ancients left this knowledge, but its keys are lost understanding of common being. In particularly, by observing the behavior and reactions of the body to stimuli protection against invading bacteria and viruses that live in the environment, it recreated a system to protect the body from infections. Notice how our system can protect itself from an internal form and react with superior intelligence. A negative source of energy is clear from outer space and reacts with the matter in some area channel of Earth and produces these reactions. The regroup these is what emerges from inside as stones. That law gave the father's legacy for our protection as an agate stone.

153- A simple sneeze

Start with an unusual secretion of mucus membranes inside the nose, as a form of watery secretion that coats the inner walls and surrounds the particles that are within their environment and imprisons its substance. Now, a violent reaction in which several muscles throughout the body contract, as a tangle of internal forces and create a reaction where the body inhale an amount of air and expands the lungs to their maximum arises within us capacity. This air is expelled outside through the nose and mouth with a huge force. We were surprised that this reaction is directed to expel the aqueous matter that arose now and that its purpose is wrapped in its adhesive intruders who are entering our respiratory system, a simple virus that has penetrated the defenses of villi created to block the entry of pathogens into our body. They stay in the mucous walls, where they are retained by the sticky qualities of this substance. Their reaction is to use the same material; They soaked into the material while it feeds to reproduce, because their system has mutated to adapt to the body's reactions. It is the law of nature survival. I recreation as an example of the reactions of the inner universe. Back to my earlier concern. What is the need for these materials have the urge to penetrate the human system in an intelligent way? Scientific advances suggest that may have been laboratory creations in chemical experiments. The reaction is accompanied by repetitions, until the viruses are expelled. If the body fails to expel the colony of infection lodges in the mucous areas of our respiratory system these continue to multiply until they can penetrate the internal respiratory system. We must ask ourselves what on earth causes this competition to penetrate within our system to these viruses or bacteria, or whatever it was in a dormant state in any wisp of dust possibly before, hoping to break off and be inhaled by any physical being that arises. What intelligence of these parasites from our body precisely directs our bloodstream? The lungs and the whole system will be invaded and colds, flu and everything related to these infections will continue. The millions who are affected annually, the cost in lives, costs millions of dollars it costs to governments in their health departments, believed covers one or more

percent of the global annual budget. And all for a simple sneeze he could not expel a stranger home. In my remarks, I created a kind of utensil to clean the cavities of the nose when sneezing occurs first and alerted us that foreign agents are present in the nostrils. Its structure is two extremes of high to low and that conforms to the cavities of the nose. Comfortable material for this area which can be used by anyone anywhere. At the same time, all kinds of drugs are suggested to impregnate and fight that way the colonization of our nose, and thus stop the infection happens to our lungs, because once it gets there no way to stop infection, which causes the body to go through a process of disease that could have been minimized or completely control. The object of pharmaceutical products based on chemicals is blocking symptoms and pain. It is to eliminate our sense of disharmony with a painkiller that separates the consequences of the disease and divert the attention of our nervous system. In any way, the progress of the disease is stopped or destroy the invading colonies. the body's natural cycle itself is used for the disease passes through its stages of progress and finished under an internal struggle that was not necessary and that chance is what made the most of the healing work. High fevers in the body are the reaction of the anti-virus that are naturally inside our bodies and engage in a struggle to get rid of infections. It is the same reaction that occurs when an injury or disharmony arises in any part of nature. His impulse is to repair and establish a state like

previous. Eventually, the details of the disease followed its natural course. Self-protection system has created a body shell of reparative influences from attacks by different infectious agents that most of us know and simply go through a process of infections that eventually the body heals itself without ingesting drugs. With the consumption of substances foreign to our system, we alter the sequence of antibody production in the future will not be present in the fight against disease. The reactions of pathogens are directed to inhibit or overcome the effects of substances that are manufactured in laboratories. The medicine should be directed to cooperate with the laws that govern nature of inner healing, not to block its effects. In the focus of the functions of the endocrine system, where the thalamus and the different lobes of the brain should focus on its principles of generators of total body functions and actions processes, mathematical statements, heliocentric theory of Aristarchus 230 a.n.e, and all the Gathered knowledge in the great schools of Alexandria in ancient times. With the destruction of the Library of Alexandria, Where They Were MOST EXTENSIVE cataloged the knowledge of the MOST advanced minds of humanity, scientific, historical, mathematical, philosophical and all branches of knowledge discovery (many) Have Been lost. We do not know if They Were stolen and hidden, or truly burned. When it was discovered this knowledge by the Christian Church and saw the growing threat hanging over her, she hastened to send a mob of fanatics to raze everything to realize the mistakes That put into practice, trying to make the deification of Jesus. sun god incarnated in person, by Saul of Tarsus WHO believe a god to the Romans and Their intention to take over the empire, of Nero. Who Doubts must seek and find. The Important part is mentally That disorientation and reciprocal approach to energy Beings to react the arguments focused on an erratic culture, misaligned forces and directs them to form a fanciful concept That bubble creations. That source from Continues to fuel the imagination of thinking Beings, began to be derailed minds distorted by paths. What the mind Becomes realities of divine energy and Its variations coagulates in a reverberation, this will continue to grow at level Such a reality is That the day will be updated every absorbed by MOST Beings to return to a new stage of maturation. Children eat into a troubled society for Their erratic concepts, Which Have That civilization will disappear matured. The legacies will manifest Themselves in the loss of civil orientation. The repetition of concepts unclarified full pro leads the imagination to create internal forces to lead Eventually That uncertainty. I have awakened from the torpor of trends manipulations created. That Should be the awakening for future civilization. Genetic manipulation, subtle cosmic energy, energy Emanations from outer space manipulated by Those for the welfare of created human systems, destabilized the dominant and controlling That forces the balance of creation. We are treading on the margin to check and manipulate energy succumbing That has made the nature of the Often planet. The imbalance of forces would lead us to what would be something like Atlantis and Lemuria. The common man has not matured a way to harmonize the energies Comprise with balance your thinking being with Variations nature of laws. Restlessness, fear, failure to Obtain a security entities promise to protect the WHO manipulation of information by the chains of transmission That have no controlling or prohibition from making baseless statements Sometimes, could result in a cataclysm unprecedented. If I mention here of these social elements, Also the fear of death is yet another incentive for Those attitudes. THEREFORE, That These statements are in this book is to alert if there is any Citizens worldwide change in living conditions, will not be the end of the planet, much less to humans. Which make a statement like that is untruthful? That no one has power. Only the eternal creator Enables us to discern how we act and Should not be afraid of Threats. By Patrick Gerril statements, author of "The Orion Prophecy" these phenomena Have Happened before and we're still here. His statements are accurate and credible Until time proves otherwise.

The inherent qualities of each separate manifestation are Transformed to the needs. According development and the pitfalls that must be overcome. In the case of a lizard, to name one example, STI development requires train engine Means to Overcome obstacles to tail to maintain a balance of movement, properties in their legs to climb on trees and, Countless Features That Have Evolved heritage and survival needs. All Beings Respond to the same needs and are adaptive evolution. The plants keep a genetic code of all the attributes of manifestation and duplicated in Their seeds, and Continues ITS survival. All that is of a different nature Creates a disharmony in the act of creation and this is Manifests Where the disease itself, the disharmony between matter and the laws, that must take ITS course. The symbols with one or two snakes observed, wound in Representing our spine Where screwed up our brains show an expression of the two opposing parts of our body and Its Contrary Energies gather in the center of the brain, and if the disease occurs they attack each other hence the destruction of internal diseases. The Egyptians, Greeks and Romans ruled left a great knowledge and scientific and medical legacy to humanity. The Foundations Have Been created in terms of the material heritage of being. Science must deepen diseases since the beginning of the evolution of being to focus Their quest to restore health. The land stages of magnetic gravitational energy are disorienting, as it reorients itself the animal species. For some reason in the same way to operate the human body is not guided by itself, for the reason. That had ancients Control of That information and practiced concentration exercises and guidance of Their bodies materials to restore health and inner Harmony with cosmic forces. It was a virtue for them the spiritual legacy towards Beings That endowed with superior knowledge. The body evolves materials and methods to combat diseases creates and stronger ways to evolve internally, Prevent the disease. I Mentioned before, it is eleven that way any species find a structure to modify, their will evolve to a higher survival in manifestation species; Have even the first creatures developed other species, derived in other different and other qualities. Of the first creatures, they Have developed other species, derived in different and other qualities. In the original environment Where all forms Emerged and varieties sharing the same facilities of water, and which by their nature evolved under surrounding conditions with each other, and ovulating and expelling them outside. Their fluids reproduction, where there were no procreation law domains, because they DNA of evolution is in the early stages organizing of development. The fact that these Were combined materials without control is the basis of diversity of creatures that populated the first manifestations of life. Climate change and storms, that Have affected, disperses over time life in the oceans and all areas of the planet, from the qualities Which emanated enabled us to Have an evolution higher than other creatures. The glands That Emerged from the combination of sodium and potassium Where the quality of attracting Emerged was created and stoked the first cell. That supplied protein for self-expression provided it with the capabilities to build Its Own Means of development and evolution. Their primary skills to Attract different conditions in the composition of the elements at its disposal. While additional cell division stoked with the energies That Were gathering with new how to generate and different energies. When agitated, neurons created with first vehicle of expression and other elements combined with That Were developed in unison in the same environment and so on more complex forms. From this center as gland provided primary neurons That shared the sunlight That provided them with the necessary elements and strength of expression to generate what after a long period Became man, endowed with a different heritage the other species That Were created in unison with being.

The way

If Beings carry out the human potential in your inside and freedoms inherited That would realize it is holding access to the Kingdom of light and dwell in it and return it to life, and return to the kingdom in His departure and change of the substance of life and death. His soul does not disappear; Changes of expression only. A state of reflection to another end and integration in purity to your destination, which is the sanctification and communion with divine substance. I have my word Receives Enters the source of my being, because I will get me as knowledge and put it into practice in His own being. Once you enter in communion with the nature of the Father, all will be revealed; Its essence will penetrate your being. The spirit soul is -not the part That Is Associated by human emotions, from the time of leaving the body and materials are integrated into the nature of God: love, compassion, mercy, and so on. The laws of creation take stage to make way mentally for a slightly more defined and elevated Processes That eats into play in the maturation understanding, and define laws and principles That mark a true definition of internal Processes, that dominate the creation of God. Those Who Do not have a comprehensive knowledge, perception of time and space and all dimensions That Being Attributed to what I perceive as the manifestations of what we call God. Inner stillness, the midpoint Where there is no conflict of any kind, the abandonment of ego, ego, being in the middle and share the flow of all that is Manifested and perceived internally. Nor is music, nor water, nor river, nor spring; When it is neither man nor a woman is: he is one. God has no sex consciousness; this is a dimension of the attributes of the human mind. To Attract the divine substance of God must bow down in the center of being and surrender to deep silence and peace. Where it leaves the I to give way to higher soul That permeates the universe, Where cease Struggles domain. It is integrated into That Dimension Where everything is one thing and the light created is confused with the material. Until That level of understanding the Teachings of Jesus rise, for I was endowed with an understanding of the nature of the Father.

161-The universal mind

What unites us as beings of creation and covers the full dissemination of what we are as beings who are aware of our existence as living beings. The universal mind that is related to our behavior and our own heritage development as individuals at the same time is part of a collective mind that is intrinsically linked to our individual self and yet unites us all manifestation. Each cell of our being, like every cell in the species, has an individual and collective intelligence rather than the laws which dominate and which responds impulse to be harmonious with its surroundings. Our body has a universe of different functions; characteristics are different from each other. What it is going to be a skin cell it is and what will manifest as hair is and so on, completing a chain or cycle. This in turn forms an inner universe which in turn is attached to what form the full manifestation or human. The attributes of being are the expression of a world that has traveled to this plane of manifestation to be known ...

Initiation of consciousness

Emotions too overcome

Faith

The term faith is generally used when a person puts his belief in something you do not understand and that references to hear from others or by what he explains is convinced of knowledge that are offered without doubt or question the source from which emanates that knowledge. Other manifestations of faith are given when one subjects another to a conviction and intended that this be put to him without explanation, as he has accepted.

Other forms of exercise this concept is the reliability you have in who produces the fact that they should believe. In religion, faith is the certainty that what is told by reference to another source is infallible and should be taken as such, without question. Faith alone is lost or defeated when it is convinced of a truth ultimately succumbs through new information that reveals a more reliable source and demonstrates the mistake of the first. Simple law of reason. The most efficient way of true faith is when self is the one who discovers the source of truth and can prove it, and all the elements that compose manifest and nothing remains hidden or that can be understood falsely. Faith emanates from truth and knowledge. If truth fails, faith crumbles. Once the custodian of a power failure reliability faith is lost and is reported to the public and nobody trusts him more. To those who hold positions and the poor use of knowledge, be assured that God himself shut the doors of heaven, who deceive humble. I am not sure that they are aware of this fact, but it is possible that something must be sure: reincarnation must continue for millions of years until they can compensate for the damage done. That is the law and they cannot escape, because within. You can create your heaven or your hell of your own nature; for that you gave you that freedom. Choose.

Suffering April 30, 2010

We are children of a Creator Father and so we recognize humans worldwide. We gave some spiritual attributes such as love, beauty, forgiveness, courage, compassion, humility and one of the most sublime after love: suffering. This is the counterpart of everything we like and appreciate good, pleasant, expression of human emotion. It is what enables us to assess the other. It is no disharmony or disease; It is only the absence of the others. This is the state where the value of other emotions takes a dimension of attracting others to make up the law. The value seizes our being and we project some energies that rise above our being and are welcomed by the divine forces of God, which in turn stems from our inner being as a source sufficient healing energies to harmonize suffering. It is a universe that wakes as nature itself that meets our world and supplies us so many emotions to catalog what practically spent most of our lives absorbing: the knowledge that we manifest. The wealth we accumulate is the reality that wakes up our environment. If we engage in various passions of the human mind and a commitment to materialism our true temporal attributes are numbed and gradually space to capture inherited realities beyond this physical plane will cloud. Interacting with the original source enables us to

express their attributes of inner maturity and adaptation to the energies that leave an impression on our intuition. The concepts set forth for generations to prepare the mind and conscience of individual beings only represent the convictions of the person who created the concept. The individual god not be encapsulated visions of one or more persons attempting to define their capabilities. The law of karma is a spiritual law that has been cataloged in this way to signify a knowledge that must be understood to capacity all seekers of spiritual elevation. It is a law of balance, the neutral point that seeks to harmonize everything in the universe. If a law is violated, the balance will seek to harmonize the status and the reaction will be the balance and cause the necessary effect for a harmony, of course, with the usual effect categorizes people as punishment is established. Initiations of spiritual maturity prepare us for greater spiritual understanding of the attributes and laws operate. We also train to operationalize vibrations that harmonize with the spiritual laws that are internalized. Learning to assess the effects of these leads to train our ability to attract internal vibrations that harmonize with them. The methods of meditation and concentration of internal forces us to rise further evolution in the spiritual plane. The fact meditates on a specific subject, a being, a situation intones us with the laws acting in the cosmos. The result is that when the positive conditions are met these will manifest themselves and have at our disposal. Meditate on a material property that we want to capture in a piece of wood, with the passing

time attracts our being the ability to execute the steps necessary to express the idea into something tangible. All these attributes of our being are part of the expressions of life that touch us the inner fiber and make us understand that there is within us a communication with something higher that can make day to day and we can share with our fellow men; see in every being a brother of creation. Possessing these attributes makes us grow and become aware of their wisdom and enables us to enter communion with him. Peace Profound Mystics and teachers who have led ancient teachings apply this term to good wishes of progress and search for knowledge and spiritual maturity which depends the soul. It is a state of inner peace that harmonizes with the subconscious and evokes pleasant memories and situations that are pleasing to our ego. The achievements of personal development, whether material or spiritual, lead us to have an inner security and project us as winners in the material sense. In the spiritual sense, it is an attribute of our soul and maturity and stillness of spirit when we overcome our sense of internal guilt and we are recognized by others. When we look at the other brothers in the same building and we entertain and understand their problems and physical and emotional issues, share them with deep love and tolerance.

It is the realization of the total being, merging with the God of creation itself. The common being cannot make or imagine this understanding because it generates more questions than faith. Only those who have crossed the portals of internal creation can become aware of this reality. The purpose of my expressions is to reach the common people to find the door and touch and they reveal what they need to know to find the kingdom to which they belong and every day more away. In the endocrine glands are the source of understanding the act created beings. In the thalamus of every created a profile of what is saved is and has been in the past; pleasant demonstrations, recognized as the best of life experiences that have lived in a previous incarnation, and part of the

mission of these memories of files stored in the memory -akásico- arises in the stages of development and maturity of the new manifestation.

The characteristics of a face of a race to which he belonged there is a suitable quality that enhances the attraction of these profiles, in addition to behavior, maturity and emotions exalting people in their -energy aura that flows from our body- are captured by this white elephant of creation. There are human and material passions being developed, but not realize his reasons. They are almost uncontrollable impulses and succumb to them in most cases, unless no other more suitable condition arises and exercising superior to our physical senses force.

166- Concept of eternal life

In the evolution of human beings from the first cell that was manifested and evolved an advanced state, a higher energy projected and exerted an influence on the interaction of this first creation. This same strength throughout the universe clothing and vibrating in all its borders and contours. A subtle force that scientists rarely aware of their existence and which is related to the very existence of the universe. An intelligent force while maintaining control creates what is manifested, as if there control all attributes of demonstrations, a subtle cloak clothing all known matter in the universe. What the human mind grasps in his temporary dwelling, unable to rise materially to other corners of the universe to meet other intelligences magnitudes whether material or emotional like ours. We live in a world subject to laws that keep us tied to the history of imposed knowledge accumulation and specialize in making this dependence is maintained. The reasons are varied and being accommodate them. Transcending concepts was an old method and sometimes present mentalities that rise to other levels of understanding and not subject to interpretations accommodative. The concepts are created pro great minds that carry an image of clarity to other beings to form a cloud of imagination itself, as mental capacity. In imposing controls the powers of interpretation, dictionaries and encyclopedias learning how to project knowledge is partly under the control of some institutions that shape knowledge to reveal, that is the accepted and practiced method to educates us. People who are not in contact with these concepts, for example, in the early stages of civilization, every culture believe their ideas as traditions which governed the territories and cultures. When meeting face to face with other more developed or less developed peoples, to confront ideas, often an amalgam of those concepts accepted is created and accepted for future generations of this mix of cultures understanding was achieved. The way these notions have matured, had to overcome so many changes, many have lost their true and original meaning. It was a practice of ancient petroglyphs schools use to translate that knowledge. They were certain that all within reach were not used. Practiced yet, fibers, leather, wood, stone, primitive metals were produced by primitive methods. What we proceed to explain in words, we seek some method of cataloging the grouping of phenomena that occur beyond our comprehension and that no human being nor science has reached to explain clearly. The human being cataloged God, all laws demonstrating beyond human understanding, that is our legacy. In this direction, it is oriented this knowledge and can be useful for any human being can address a more personal knowledge of creation and concepts related.

What it is taken for granted that it is the God of every human being who accepts this designation for their spiritual strength and soul and existence in this earthly plane. The way in which this knowledge should be interpreted as the staunchest and fervent processors of these original concepts introduced. The union of civilizations, migration of ethnic groups from the northern steppes who brought their cultures and concepts matured for centuries. The migration of the inhabitants of those regions, was one of survival by climate disasters, continuous wars between peoples, food shortages and survival of races. The largest of its resources was the livestock and agriculture, besides fishing. Once this was reduced by the harshness of the weather conditions they were becoming nomads seeking new lands and territories to settle. That's part of the civilization that arose from the way the people survived. This is a summary of how to give an idea of the migrations of the early days. The maturation of scenarios beliefs and concept formation of one God, was the best way to express what the eyes and experience happening every day to people who did not have a clear understanding of natural laws. He lived daily repetition of the births of new creatures, curiosity to raise their minds to the causes recurring daily around. New creatures are recreated in front of them and their care were developed with new and higher powers of expression. The acculturation of peoples, added broader ways of relating, raised the imagination that has reached our era by way of contact with that knowledge.

That link history some of these concepts on how to know the inner picture of what was lost meant the inner form of cover and publicize these legacies and the maturation of the inner purity of those original beings. The elevation of its inner forces to communicate their emotional of what for them is their way of visualizing the divinity idea.

At some stage of human relations concepts, they must have matured and become part of the story which was published for periods of time. At that stage of magic evolution, of those original versions broke. At some time in creating a concept of a supreme being or universal law which was all content matured, the repetition of the laws of creation and the constant improvement of the phenomena of creation, created in the human mind an intrigue of how things work.

The reason for the phenomena that are repeated, the creations of nature that is repetitive, that does not go away, every time your life is renewed. Curiosity manifest that things never seen before the astonished gaze of the only ones who realize these facts. They can interpret in their natural environment and keep it in your memory as something that must evaluate new ways of cataloging the events to their way of life. The maturity of these mental pictures is what give us the certainty that somehow the concepts we have before our life could mean different things. The heart of this question lies the potential changes of what emerged as a reality in the minds of those who developed the concept and what then interpret them. The question arises lying or decomposed concept of reality. To be replaced with what is true. Arises the concept of guilt of beings. What he proposes someone as a reality of truth is the opposite to thoughts or mental maps of other souls that coexist in the same environment. The darkness of clothes concepts of mind, which cannot overcome this stage of life must accept theirs and others' faults, and born the concept of sin. Feeling bad because another be it projected an opposing view to his and that by acting outside its scope brings him being branded a sinner, liar, etc. The form of Christianity uses derogatory phrases to bind attitudes or beliefs of beings who do not share in their enforcement all the atheists, contrary to Theo's they preach and created laws. In my opinion Atheists do not exist, they are only human beings who do not accept the way ideas that for a long verging on blind fanaticism projections.

If it could recreate the universal history in its proper historical context, as they occurred, every fact, every reason, every detail, take a sample of any time in history and to be able to recreate outside the realities counted, we would be surprised of things who they were left off the books. These observations are based on the curiosity of things that are discovered daily and slopes that overlook reality. Just feel an inner emotion upon hearing a word intonation, a whisper, a musical note, a vibration that surrounds us, can penetrate our conscious being and moving into an unknown world of ideas and truths that sometimes words cannot bring to reality. Glimpse the passion to penetrate and transcend the other side of the imagination, captured in an imaginary cloud where they could recreate a strictly belonging to a link of creating energy, would bring to reality a power to discover that cosmic area of the mind where being with non-being, and to add a remote knowledge of the common mind communicates. Many developed beings have achieved this development and managed to transcend, or travel outside the usual energy materials and leave their personal achievements and aspirations, to capture that reality. Just feel that what has been achieved would ruin of his discoveries. The notion of something inside places on the border of life and death, both physical and mental. The terrors seize accumulated by centuries of neurons that process the survival of the matter, which succumbs to pain and anxiety of being another victim.

Darkness November 11, 2010, is defined as the absence of light. For active mind that values the consequences of what could be felt physically and deny or not attribute an active role in many phenomena for which no one has a complete definition, which believes in our understanding a motion image to which you we devote efforts to find their real qualities of existence. I take time meditating on the qualities or qualities of darkness. Although it is not something tangible, it's a vacuum of qualities that other active causes must move to exist. Where light does not penetrate, conditions are appropriate for reactions that would not be activated if it were not for the absence of light are given. Although this is a failed assertion, it strikes me that, although not stated and studied law is a quality that motivates different reactions. In absolute darkness, perhaps in the depths of the caves where no light penetrates living conditions and creating species, plants that develop with the minimum amount of energy is created.

This shows that where no light penetrates a fully reactions are only possible when the darkness is present are given. Even the ability to view and creating organs that develop species adapt to darkness; or whether, in the absence of light, matter it acquires other unexpected features that put us to think of other phenomena of creation itself. A reduction of light energy or rebounds of these phenomena which could not occur unless some degree of darkness will hinder the full and constant penetration of many reactions that occur in the material itself. A seed constantly exposed to the sun does not react the same way when under the earth and darkness only allows you to be reached by the amount of light that gives the right combination for the emergence of life on this plant. Of course, other conditions are combined in these reactions. Darkness is a condition that makes your counterpart, light, appears. In our physical reactions, darkness has a protection factor that alerts us and makes us react to avoid mishaps and dodge objects. The simple movement of a shadow causes us to react to hazards, sensuously detect the event as follows and are activated in our system of protection necessary to react appropriately alerts. Without these phenomena of shadows created by the darkness would not have developed in our physical system these complex reactions. It has been discovered in our endocrine system and brain areas of neurons that respond to the call given by fear or the ability to react appropriately to dangerous situations. This is a condition that is part of our survival system

natural, which was developed in our system for survival from the beginning of creation. If dark conditions, the way we react completely changes; areas of our system that are not common to light are activated. This shows that darkness is an important factor in everything around us factor and everything that has been created since the beginning of our existence. We could imagine under what conditions human beings and all created creatures have stated and evolution would have been quite another. The brain, which is a hotbed of chemical and physical reactions, where the current is handled is such potential that the mere fact of any disturbance to these functions creates what we classify epilepsy. This is the case where large amounts of energy are released outside their usual conductive channel and the reactions are different. It is a download entering other areas where it was not intended and makes our system central nervous enter a crosstalk and downloads that are diverted to other neural stem resulting in many cases reactions are convulsions of the muscular system. By the mere fact of receiving these discharges muscles contract, per the capacity of energy that invade. Under these conditions happens sea of experiences of all types, such as hallucinations and euphoric states where they live and feel different emotions that sometimes lead to the beings who have them enter a mental state of receiving orders or commands to do things that were not perceived. Amounts of stories have been written of this phenomenon. During evolution, they have been rescued stories of characters who have distinguished themselves in their performances to be epileptic and usually known cases point to religious reactions.

The shape of the brain to rest and regain control of internal powers, block our conscious part and penetrate the subconscious to enter a more comprehensive regeneration process. You must do this in a peaceful climate, noise-free, and that all processes are in harmony. It is precisely where darkness should play an important role. For my observations, it is a period where the reactions to light are minimal and this makes internal processes as appropriate in the absence of as much natural light. If we process elements that invade us from cosmic emanations, it would be logical to think that the body and its components have been able to receive and process certain rates of these emanations. If disturbances that decrease or increase these radiations, the body must regulate their receptive system to compensate for these changes to protect their internal processes arise states.

Renowned scientists in the world give opinions that are credible for most of his colleagues. They claim the nonexistence of an energetic substance; this substance does exist-light and darkness; sodium and potassium. What it manifests the divine creation of being, a profound reason that only try to explain God with materialistic theories and the creation of the universe. Whatever the evolution of the being and attributes it is given to the inherited human intelligence to recreate a divine world. The continuous evolution, following cosmic laws of creation for material physical parts in being, does not detract from that divine spark that emissions associated with sources that are not noticeable or have been explained by science so far. The entire range of human emotions that pervade the reality that accesses every human being and enable him to experience them through their existence in this earthly plane and then completed his period of manifestation in a treasury body reincarnated in a new body for follow a spiritual evolution to achieve a balance and greater spirituality. Human emotions not be circumvented and yes reached in accordance with the intonation of our divine inner self that we attract to us voluntarily and involuntarily, and we project of ourselves to another as a harmony of the own energetic qualities we use to shape our expressions and are an achievement of our individual personality. These emotions we do through our life and we can mature as our intellect are not covered by any religious teaching, it is only an attribute of the soul-personality individually. we can instill ideas of their potential and how to achieve it. Only the individual has the power to access it. It is a proven fact that the resurrection and reincarnation follow the same laws by which human history has progressed. It is also something real that the projection of being through time-space is possible. The return of the soul in a new body is the same process. Jesus taught them this truth to the Jews and apostles that to enter the Kingdom we must become children again. Resurrection practices include great relationships in the history of religions. Peter Resurrects Tabitha -Dorcas- female apostle priestess. It is repeated here a story of an apostle resuscitating a follower to make known the mystery of the resurrection. The same description of Jesus and Lazarus.

Reincarnation

Principle of eternal live

The term refers to a process dating back to repeat the incarnation or manifestation of the new attributes existence of an entity that lived in a physical body in an earlier period relative. This word was known in the ancient teachings in the Old and New Testaments. Attribute that was observed by ancient civilizations and the only documented were the Egyptians. She was adopted as a superior knowledge, but they could not decipher everything, which was declared anathema in 553 A.D. The form of resurrection was adopted to change the tone or meaning of the teaching of reincarnation to which he applied the word. At the Council of Constantinople, in 553, by order of Emperor Justinian, he was declared anathema and heresy. And the concept was removed from the Catholic-Christian teachings. He is failing to give a logical explanation and be a Coptic term and Greek put the authorities face an existential reality that challenged their interpretation of the word and, at the same time, the process of resurrection taken to introduce the Christ of Saul in Pauline writings. Statements and put into records of the resurrection of Jesus and returning the area where they are the dead was for them a threat is brandished like a sword over their interpretations from Saul's used to justify their interpretations and personal preaching outside Jerusalem on the resurrection of Jesus. The term resurrection includes only the phase where the soul leaves the body freely or be mastered control that power, which was common in the time of Jesus and was part of his teachings. In addition, it was known centuries before and used in many writings and in the same Bible and the writings of Qumran.

Resurrection

It can be described as the attribute to discard the physical body, either voluntarily or involuntarily, and consciously return to the same physical body – astral trip. This is the understanding accepted by the old religions practiced and taught in the school of mysteries. No one had a definite teaching on these natural phenomena summarized in the attributes of being created; one's soul or spiritual energy, leaves the body for short periods of time. For teachers of different civilizations which aims to dispose of their knowledge, it is a practice of persuasion, evoking the mysteries of his spiritual attributes. These and many Avatars in different times, have tested their conditions of spiritual advancement through physical manifestations. This would be a direct evidence of the causes referred to religions attribute to seek spiritual leaders claimed on each sect. The purpose is the same in each country or race as a spiritual leader proclaimed. Asceticism, purification of human matter, the domain of energy creation on the subject, gives a very different reality to what is commonly known. Spiritual elevation is what everyone yearns to understand inwardly. It is not a simple matter of desire, of feel. Many ancient in different cultures, civilizations took steps to induce his followers, by using hallucinogenic substance to give that sense of elevation to a state of higher consciousness. History is witness to those times, that those ways to conquer the minds of the unwary were devised, they are symbols of history that dwells in circles and civilizations succumbed to such practices. Behind the written history are witnesses and key players in that frame.

The release of spiritual energies is not an alien phenomenon to the reality of the physical laws of the human body, or consciousness of which we are endowed by creation itself. They are additional conditions to what we are natural to our being. If this feeling is experienced at times during sleep and wakes the body relaxed and notions of new knowledge. The entry of soul energy in the body is a mystery to many researchers. They are surprised that the subtle energy of the soul is simply a vibrating spiritual energy. A cluster of imperceptible vibrations that travel through the universe like a wisp of projection of the source that contains everything. This contains attributes are the characteristics attracted to this earthly plane by the laws of astrophysics. On planet earth where they inhabit the bodies of human beings, by adding everything that makes up this planet in all its manifestations is the correlation of the energies that are attracted or planned to spheres of manifestation, in accordance with the stages maturation of energy and the laws of attraction to create. The resurrection is a release of a physical being of an amount of energy that enables the functions of a body. The mass containing a body, accumulates, comparable to the energy of matter which comprises vibratory rate. Based on the molecular structures and atomic components, this mass has a magnetic field. This in turn, is a mass of energy that attracts and expels of its components, waves line with other materials that are expressed in the universe. Emissions laws must be in unison creating a tuning fork of attraction and repulsion in the spectrum of cosmic energy. The composition of the physical universe, like everything that exists, due to the influences of materials and the laws that dominates.

Science as such only displays captured by the human mind through the instruments used to guide the understanding of these laws at a level of human capacities effects. We are the only ones who realize it. For more than penetrate the understanding of cosmic phenomena, the human mind will continue to grow in stages. Data accumulation leads to a clear vision of the phenomenon and laws that interact with what life is cataloged on this earthly plane. Possible awareness of this reality is an escape of pure energy into a crude reality of nature. Traveling the confines of its purity to enter a stage of introspection of its reality of destruction or degradation that allows you to manifest and become known outside their environment purity. Incongruity of the human mind to understand its nature. Being a god power that has never been surpassed by its being deceived by others project realities. History will be the teacher, within its confines is the story that never came to light. Nobody can hide a lamp on a high mountain, without others to see, from the plains of Genezaret lake temple, the historical high mountain. Whoever finds meaning to these words, not like death, it will return to life and no one can clear his conscience, she accompanies him forever in his travels back and back. There will come a moment of understanding of the sources of truth that embraces the universe. It will be like a temporary illusion of mental reality of understanding. The accumulating data and experience for eons of time will realize that the concepts are elusive, which is now a law tested within a short time, will turn his own dating that holds in the annals accumulated evidence.

Reincarnate

When he finishes his stage of life here on Earth and rejoins the universal part of divine essence, no matter the degree of progress that has been reached on this earthly plane. A simple change of state of manifestation is cataloged death. The mere fact of having to define these concepts, and when they were confronted with evidence that were already known and applied in different cultures, exposed teachings that were based on an interpretation of history that does not tolerate scientific evidence.

The Kingdom of Light The physical part of our inner realm is equipped with a web of neural networks that simultaneously are internal censors which establishes a communications network where traveling million impulses to the remotest parts of our physical structure. Its function is to receive information or signals that enable constant communication is established and carries instructions to the different areas of control to the materials necessary for the proper development of our bodies. Thus, if there is any nerve exposure, the signals will be sent to areas where they are connected to alert the harm suffered or the material needed to restore a specific function.

The rebuilding process will break loose immediately. This center is on two levels of connections, internal and external, with the space outside our physical spectrum material. Creation gland receives all impulses from the cosmic part and harmonizes in the scheme of the physical manifestation of all creation emanating from the cosmic mind. It is just a link in the unknown. Physical pain is a feature that causes any injury affecting a nerve terminal; It is one of the sensations more pronounced in our being. Intelligence manifested in these reactions not be exclusively of material nature, since the signals will be sent to areas where they are connected to alert the harm suffered or the material needed to restore a function. Sending the necessary impetus for genetically engineered cells and other aggregates are believed to repair the damage with the same characteristics as the previous one. Why? Realizing these attributes is where emotions of creation emerge. Our emotional system that overwhelms harmony. They believe it or not, that we create harmony in the way we react to these impulses. We are endowed with a resistance to pain or any emotion that we impound. Many succumb to these experiences, but others are trained to resist. What mystery surrounding this attribute within us that makes us aware and sensitive of all our powers of expression? This attribute should be tied to an energy that gives us life itself and must be part of the vital energy flowing from the cosmic God to our material body and spiritual part. The reason why I express this is because once life ends this energy leaves the body and feelings disappear altogether, like all the other attributes that distinguish us as created beings. The need to attract energy from sources that give us the qualities interact is part of that energy that vanish instantly. The secrets of the ancient Egyptians have been rescued by great enlightened who quietly have been dedicated to collecting these teachings and secretly for over two thousand years, they have kept the secret with the promise of making it known to those who are deserving of that illumination. Those interested should search the initiatory and esoteric schools, as they did in ancient minds seeking enlightenment; simply typed on their computers and the doors will open. Lamas from Tibet started in China, India gurus and San Jose, California, Mexico and many other places worldwide. In the era of the Egyptians, the people went in search of this knowledge and traveled great distances to get in touch with schools and be admitted. Where did all these instructions emanate are faithful copies of an earlier demonstration? Why plants are duplicated in their seeds, human beings in their evolution? - What role does our imagination and power to attract the cosmic to the physical and vice versa planes? When these powers cease? - What is this mass of energy that went and where it goes? - What law regulates the period between a phenomenon and another? The reorganization of matter in so many varieties of expression that surprises us every day, diversity manifested. The common mind is not aware of the surprising fluctuations in the phenomena occurring at every moment. Others are a big circle of expectation and everything that happens will be a new dawn in your life. The mysteries that encloses the evolution of species over millions of years, keep us busy, scientists, thinkers, mystics, and all that branch of knowledge that seeks to unravel the evolution of species, the human knowledge largest chain, which is the bearer of its own information and that could not decipher its contents so far. Any form or explanation end forming another link in the chain that simply expanding complements the way it is projected. From the cosmic understanding to the lowest of what is seen by the human mind expanding. Each being and mind is a world of individual creation. All that is processed to individual conception, and will be part of a personal world; in turn, will form part of the flow of universal intelligence; It may not be rejected by the original source. That source of human projection swells the universe itself in its expansion. In other words, if there is this manifestation of intelligence, they will manifest themselves to accumulate the expression of the human soul and its subtle emotions, which is what can be exported to the components of the original creation. Intrinsically a world of possibilities for interpretation of the original idea opens, but universal concepts of sharing the same idea. It will be influenced by the generality of knowledge achieved. Knowledge that was saved in all possible ways: in stone, scrolls, secret keys, caves, writing, and so what we have today is how little has survived the ravages of nature and humanity. In the first historical data of the resurrection Elijah the prophet living in a cave for some reason the first example of resurrection shown in documented history of mankind is mentioned. Elijah raises a boy, son of a widow. Again, the resurrection where a cave is part of the picture is reported. When a hidden truth declares a symbol, no matter the past few years: the symbol remains the symbolism that was introduced as teaching, and the key to its original interpretation is not lost. Once manifested in an expression of imagination, powers invade the universe, but the immediate truth is extinguished as the flame. With practice and spiritual evolution, we prepare for that every day we go deeper into becoming aware of the divine nature. All that being created goes through a process of visualization and awareness of how to transform is taken, transmute an idea into a real demonstration material. Everything that exists has materialized in this physical form. From imagination to creation, which they are attributes of human beings. Thus, being seizes the qualities and energies of the thing to manifest and plasma into a reality. That is part of the cosmic laws materialize into something real. The man has been consciously or unconsciously created for that. Imagination, the source of the internal creation, is the inner reaction entire range of energies that make up a controlled and driven towards the harmonization of own energies that manifest themselves in a creative

impulse and the evocation of a desire current that it will take shape and reality. No need to give birth to at the time, this inner force will continue to act and bringing the attributes of the thing until the human being want to physically carry out the plan imagined. It is both mental-emotional form and physical form of
expression. The seed will keep its essence creates until conditions are right manifestation to give a good result, and if these are manifested in the object is encoded in its genetic code. This process is identified with reproduction in all nature, following the same laws. In the cell, the regeneration process produces cells that will be programmed for a specific purpose and specific functions in the areas which, by their identification, the internal system, will be taken to the right place to which they have been designed. The command center in humans is run by the set of glands led by the thalamus, pineal gland, neurons, brain membranes, central nervous system, the lobes of matter in the brain and the entire range of materials and components belonging to this old elephant living in the place of creation. This is the grail of creation, the mystery of mysteries, the Incarnate Word, where the first cosmic vibration of creation is recreated, the divine logos, the word lost. From the first time this gland re-created the evolution in vibration and took harmony in your environment, created for himself a profile and a memory of all components and was created during its development all the qualities that were necessary for progress. If your home depends on the evolution of species, it must have arisen in unison in different manifestations and divisions which in turn kept the information as the seed, and the next demonstration would follow the same law of manifestation. The fact that science point in their recent findings about the qualities of this control

center within us and the attributes that have been proven compared to the ancient knowledge that dominated some of the enlightened of our planet. The first cell that arises and survives and was encapsulated in its environment due to inclement to which he had to face, gave him the attributes that Father-Mother Creator said in his involution was the material, created from a spiritual space projection the same subject. The concentration and meditation exercises show that this center reacts to the concentration of energies that we plan and headed to the centers that compose it. The same when we focus on it and let it take control of our communication centers, we noticed how an inner harmony and energies of the spiritual-emotional travel through our body from heaven to earth and earth to heaven is established. Here's the interesting thing about this meditation. Much of the knowledge that has been accumulated for centuries and civilizations attributed to different states of spiritual elevation that is not explained his actions and is still under investigation at present. Events seizures are attributed to states uncontrolled areas of the cerebral system where the glands that make up the system are located and suffer downloads uncontrolled energy is supposed to flow normally for proper functioning of the sequence in humans and those who dominate scientific knowledge is cataloged as rational beings functional. Within the parameters set by current science of human behavior and mental health parameters appropriate to what usually conforms to the patterns of the manifestations of human beings themselves conduct are established. Based on this premise, it should be noted that because of events that have been known since ancient times and have accumulated in the annals of ancient civilizations such as Egypt and others who excel in spiritual progress worldwide who have bequeathed a cultural background for humanity, must analyze this fact more clearly and carefully. While it is true that these energies uncontrolled phenomena cause unusual features, the fact remains that because of them a reality that has never been considered from that scenario appears. The capacity of human beings to have a wealth of unknown functions and that science has tried to explain, and which now has been an advance directed towards this area of science. The ancient Egyptians mastered this knowledge and used it for their advances in all aspects of his life. In the regions of Greece, India, China and ancient Mesopotamia secret domain knowledge of this energy is transmitted. The secrets of the Library of Alexandria and many of the old files kept this secret knowledge, most of which was destroyed. It's the same or like the proclamations of the Council of Nicaea, that after death and religious persecution most of this knowledge disappears from the known world stage. The secret, secrecy must be the place to keep this knowledge away from religious functions. That has been the experience of black history we have inherited. Control of this type of natural energy in humans and is a spiritual heritage of our divine creation. As I said, control of these energies is ours and with proper discipline and knowledge attribute can be applied in the spiritual advancement of people. The activity of the institutions dedicated to creating causes fans today remains active, trying to keep their domains and ensuring that the truth is not known. Current knowledge with truth, is forbidden, just not be sustained only by the declaration of being elected without a real reason. Such statements are valid for their councils and pawns on the chess of your board. Human intelligence is superior and has crossed oceans and cultures to unravel the truth, that we embarked with models of a civilization that crossed the Atlantic with those cramps and bruises of what the imagination of the cloisters, organized for gross this part of the world. Knowledge takes possession of human reason and not be stopped with threats or persecution. In the old schools that were called, of the mysteries were taught and currently metaphysical processes that have been called to catalog some understandable as a branch of knowledge taught. Calling mystery school is not a designation of rarity or something unusual, it's just a way to catalog a scientific knowledge that is not at the level of ordinary people, and it was studied by those who had advance education and training for the advancement of people. The search of the nature of God in our being and understanding of their true spiritual being in our soul must be the goal of all created beings. To a greater or lesser extent, will free us from ignorance that has taught us for centuries, path that motivate the search for inner truth, liberty dwells there, individually and collectively. Before these unconscionable mismatches in the old schools met the great enlightened to know (Jesus, Eratosthenes of Siena-Alpha- Beta, Copernicus, and the sequel of brilliant minds we study today.

186-Breaking created myths

Daniel-Rops, Jesus in his son temps

You agree with this brief chronology of recent days lived by Jesus - Thursday, April 6: Dinner (at sunset), detention Olives; - Friday, April 7 (night) process, crucifixion, death; - Saturday, April 8: stay in the grave; - Sunday, April 9: Resurrection (at dawn).

We will now carefully study the claims of the Christian tradition, and to make their criticism. Those who drafted the centuries IV and V the Synoptic Gospels, the apocryphal John and did not have all the necessary elements to perform an unassailable work. Lacking communications, easy to consult libraries, epistolary relationships as comfortable as in our time, it was very difficult, if not impossible, to carry out a perfectly synchronized work. At that time, given our current methods of verification and control, he was not wanted fabulist who also were not even Jews.

His super abundantly mistakes prove

Customs and Jewish rites do not know everything, far from it. Here we will reproduce as relevant critical analysis of Auguste Hollard, in his Origines des Fetes Chrétiennes: "The last meal that Jesus took with his disciples on Thursday, the eve of his death, left in the memory of these an indelible impression; It was then, for the last time his beloved Master gave thanks, broke bread while, then distribute it as a symbol of union, and when filled the cup and blessed, before passing it to his disciples. "

There was nothing there that was not perfectly in accordance with Jewish custom, and to the formulas of blessings, which read as follows: ". Blessed are you Lord our God, King of the Universe, which do produce bread to earth" and "Blessed are you, lord our God, King of the World, which has created the vineyard". It was during this meal when Jesus told his disciples: "I will not drink of the fruit of the vine until I drink it new in the kingdom of God." (Mark, 14, 25.)

That's where it should take place the next meeting, hence there will be no opportunity or time to meet, because the Kingdom is coming. If Jesus has the feeling that, before inaugurate it, must pass through death, moreover, it is not sure. A few moments later, in the garden of Gethsemane, ask God to save this supreme test. "So, Jesus had been unable to think of founding, apropos of that last meal and in commemoration of his death, an" institution of the Supper "that in any case the imminent prospect of a celestial appointment would have done well superfine.

The last supper of Jesus is not of any of the characters of the Passover meal, if not the final hymn (Mark, 14, 26 and Matthew, 26, 30) which, in any case, could designate the Hallel "Talmudic Discussion the liturgical use of Psalms 113-118 focuses on how the Psalms incorporate gratitude for past acts?? God of salvation and confidence in the future redemption of Israel of God "

But not in her or bitter herbs, and the four cups, even the paschal lamb, which would have symbolized Christ better than any other food item, nor the unleavened bread, but ordinary bread (Arton in Greek). '

Mark (14, 22-23) and Matthew (26, 26-27) we read: "While they ate. Jesus took bread, and blessing, broke it and gave it to them, saying, "Take, this is my body." Then taking a drink, after giving thanks, he gave it. " To view this meal a Passover meal, though it seems little- have to admit that cup of blessing which follows the distribution of bread was the third of the Jewish Passover ritual.

Lucas was more prescient and did start the meal (22, 17) with the blessing of the cup. He not puts him "as they were eating," which effectively disrupts the order of the food, and just the meal with a glass distribution, which could, in extreme cases, be very well the fourth ritual. " (Cf. Guignebert Jesus.)

But we still expect other contradictions. How can we accept such absurdities from eyewitnesses, as John and Matthew, and that ignorance of the traditional, so punctilious, from pious Jews and Jewish ritual Luke and Mark '? For the Synoptics, that is, to Matthew, Mark and Luke.

Jesus celebrated the annual Passover before his execution, and handed them bread and wine, changed into flesh and blood mystical. For John, however, it was at the time the Passover, which rituals lambs were immolated in the Temple, whose blood would stain the altar (animals that parents were brought then home for preparing will consume a family, per a very specific ritual), at that moment it was when, for obvious esoteric symbolism, did expire Jesus on the cross.

Well, we have an obvious contradiction. For the Synoptics, the night before the day of execution on Golgotha, Jesus instituted the Supper, during his disciples. It so happened Thursday night, and as, per Jewish law, the day begins at sunset, was already the beginning of the 15th day of Nisan. During that day, it was when they had to sacrifice on the Temple Passover lambs. It was during the night that followed immediately when Jesus was arrested in the Garden of Olives, when he was tried and executed; therefore, it was the next day, i.e. Friday. Then he went into the tomb on Saturday and was resurrected on Sunday morning.

On the contrary, per John's account it was obviously a snack, a meal, and the episode of bread dipped in wine and offered to Judas it is proof of that. What does not say is that it was an institution of the Supper, a Passover meal or, in the ritual and

Judaic sense of the term. Jesus' arrest also occurred the night of the 15th, but Thursday night of Nisan 14. The next morning, the Jews did not enter the Roman
Praetorium for fear defiled and unable to eat at night the Passover lamb. (Cf. John 18:28.) And, therefore, it is the time when those lambs are sacrificed in the Temple, thousands, when Jesus expires on the cross.

We are at noon on Nisan 14. There are therefore two days apart with the Synoptic. And yet, these events, oh miracle! fall on the same days of the week: on Friday, the execution took place, and Sunday resurrection. The meaning of these special effects is unclear. Because Friday is the day of Venus, aka Lucifer, and Jesus expires on the day of his Adversary. Hence the ban, for centuries, to celebrate the Eucharistic Supper dishes or cups that have copper in its composition, because this is the Venusian and Luciferian metal.

On Saturday, the Sabbath day of rest, is the day that happens in the silence of the tomb. And on Sunday, the day of sun, light, takes place at dawn resurrection. Who wants to try too many things, does not prove, says popular wisdom? The events, as the count Matthew, Mark and Luke Synoptics, leading to impossible anachronisms to admit, and show that the anonymous who wrote our Gospels in the IV and V centuries ignored the most elementary logic. If not, how admit that the first day of Passover, which must necessarily be devoted to rest, as inviolable as the Sabbath (Exodus, 12, 16), in a week that was a real spiritual "retreat" (pp. Cit., 12), they could occur mount the arrest of Jesus, deliberation of the accusers each other, and then with Pontius Pilate, the purchase of a painting by Joseph of Arimathea, and burial of Jesus?

In his Chronicle Pascale (initium), the ancient author Apolinar pointed out, rightly, that an execution in Jerusalem as sacred as the 15th of Nisan (April) day would have desecrated the Passover feast was prepared, and could have triggered an uprising over the Jewish masses. Rome, which was very prudent in these sensitive points, which had agreed to withdraw and hide the badges of his legions during his stay in Jerusalem, which had removed the shields of gold for the temple for being offered by uncircumcised, this Rome, which had shown many times its respect for the Jewish religion, it would not launch such a court challenge. Moreover, the Jews could hardly have been dispensed to attend the ordeal, they who (per the Gospels) Pilate had requested the arrest of Jesus. But the law says Passover explicitly: "[on that day] you do not occupy of any work." (Numbers 28:18.) During these holy days, Jerusalem was invaded by thousands of pilgrims.

Never the Roman Praetorian and the Jewish Sanhedrin could have come in that day to the trial of Jesus. When, some years later, Simon Peter also will be held during the Easter week (another uprising more), Herod Agrippa take care to postpone his trial for "after Easter". (Acts of the Apostles, 12.4.)

In addition, the Synoptics themselves confirm to us that the arrest and subsequent trial could not take place these days: "They (the chief priests and scribes) said:" Do not be at the party, do not go to riot the people. "(Mark 14: 2 and Matthew, 26, 5.)

Other than that, the interrogation of Jesus during the Passover night was impossible legally, and we know how the Pharisees and the teachers of the law to these subtleties and girded those legal taboos.

Indeed, in a city without night lighting, which, like all ancient cities, had a Draconian subterfuge (to alleviate the fire), it was physically impossible to gather immediately after the arrest of Jesus, and toward one of early morning, a whole Sanhedrin, consisting of seventy-two members, all elderly, the heads of the Cohanim, the scribes, the elders of the people and the numerous witnesses. Furthermore, per the law, the Sanhedrin, to judge in criminal matters, 50/0 could meet day and night ever "because the darkness clouding the judgment of man."

Moreover, in criminal matters, when the guilt of the accused was recognized, the verdict could not be taken until the next day. Therefore, the law, "criminal prosecution could not ever start the day before the weekly Sabbath, or the eve of a religious festival" (cf. Michna, Sanhedrin IV, Babylonian Talmud, p.32). And there's more: it was not possible that 15 Nisan, analogous to compulsory rest day a Sabbath. Simon of Cyrene "came from the field," where have been working (Mark, 15, 21, and Numbers, 28, 18), not to be forced to help Jesus carry the cross, as this would have been a job. Finally, the output of Jesus, followed by his disciples after the Passover (or the "alleged" Passover meal) food, described in Mark (14, 26), is incompatible with the formal requirement of Exodus (12, 22), which prohibits outright leave the house where the Passover meal takes place until the next morning: "Let none of you shall go out of the door of the house until the morning ..." (£; COAO, 12.22). on the streets of Jerusalem could not have, wandering around, but the Roman patrols, who were watching for a new uprising came to not disturb the party. And every Jew (easily recognizable by their typical customs) had been arrested on suspicion unerringly.

They are now a series of improbable things and apparent contradictions. The main reason that justified the arrest of Jesus was that claimed to be king. That would result in the inscription that Pilate himself drafted and sent nailing, by use of the time, above the patibular cross. And that was what the prosecutor reproached him during his interrogation, and that Jesus did not deny (Mark 15.2). Well, that is known as the crime of rebellion. And to be with Jesus, surrounded by his men, all armed with swords he had recommended them to be ready, if need be at the expense of selling their cloaks (Luke, 22, 36), Pilate orders a true armed expedition, comprising a cohort, i.e., six hundred veterans, elite soldiers commanded by a tribune, military judge with the rank of consul (John, 18, 3 and 12).

The contingent of armed Levites the Sanhedrin adds that little Roman army is not there but to manifest the loyalty of official Judaism. Everything seems therefore assume that, when Pilate who ordered that judicial expedition, he will take Jesus who once captured. Well then, that's nothing! Jesus, according to the anonymous writers of our Gospels, will be brought before the Jewish religious authorities, and the whole process will focus, in fact, on a charge of blasphemy. At each end, he could have sustained the hypothesis that was conducted before Herod Antipas, as this is the Tetrarch of Galilee and Perea, and to represent him their temporal power, legitimized by the agreement with Rome. Herod Antipas was precisely in Jerusalem at the time, in his palace, and Jesus, being a Galilean, depended on his authority. But our Gospels tell us that Jesus was led first: a) to "Caiaphas, the high priest" (Matthew 26.57); b) to "the high priest" (Mark 14.53); c) to "the high priest" (Luke, 22, 54); d) to "Anas, because he was father in law of Caiaphas, who was high priest that year ..."

(John 18, 13). In the end, to whom Jesus appeared first? ¿First Anas or before Caiaphas? And Daniel-Rops notes with pregnancy: "The annoying thing is that the text of the Fourth Gospel is very confusing at this point. We read that first led Jesus to Anna's, the father in law of Caiaphas, "high priest that year" (18, 13). Then comes a scene of interrogation, followed by the denial of Peter, who seems to be the same as the Synoptic located in Caiaphas; Then verse 24 states: "Anna's sent him bound to Caiaphas, the high priest." To achieve the logical sequence and both the agreement with the Synoptics, should we put verse 24 after verses 13 and 14, place which, incidentally, occupies an old Syriac manuscript and in Cyril of Alexandria. But then no word of what Anas said Jesus! "(DanielRops, Jesus son temps, p. 496.) In fact, it is known, and involuntarily, a few pages later (p. 501) Daniel Rops shows that during interrogation the pontiff said Israel could not lift in any way an accusation of blasphemy against Jesus. For this reason, we, for our part, in the event of the appearance of Jesus before the Sanhedrin see a sequence invented by anonymous scribes of the fourth century, who, being Greek and Semitic, tried to liberate Rome from the responsibility of the Jesus' death.

Now, Christianity was the official religion of the Roman Empire, and at all costs had to deal with punches to the imperial power. Instead, it is quite possible that Jesus was led first to present Tetrarch, since Herod represented the temporal power Judaic, while Roman Pilate represented the temporal power, occupying and protecting power, and therefore superior. And, once again, the charge raised against Jesus is to be claimed king. We have the proof in this passage associated with the above activities of Jesus: "The same day came some

Pharisees to say," Go away, leave here, for Herod wants to kill you. " He replied, "Go tell that fox ..." "(Luke, 13, 31.) Why did Herod Antipas, Tetrarch of Galilee and Perea, and at that time wanted to kill Jesus? Because the latter represented the Davidic and royal legitimacy, after his father Judas of Gamala, and declare it to be pretended king.

If not, what came that hatred of Tetrarch? What could give him some lessons of piety and collective moral taught the people? What could offend him the Gospel message intended? Anyway, the fact is that Jesus appeared before him after his arrest, and the story that make us about it contradicts the precedent:

"It Hearing of Galilee, Pilate asked whether the man was a Galilean, and having learned that he was from the jurisdiction of Herod, he sent it, who was also in Jerusalem in those days." When Herod saw Jesus, was very glad because had long wished to see him, because he had heard of him, and hoped to see him do some miracle. He gave a lot of questions, but Jesus said nothing. They were present the chief priests and scribes who accused him violently. Herod, with his escort, treated him with contempt and, after they had mocked him, having dressed in a shining robe, he returned to Pilate. On that day, Pilate and Herod became friends, for they were before enemies ... "(Luke, from 23.6 to 12.) Now, he says Daniel-Rops, a large part of the commentators estimate that the garment was a white robe analogous to the military tribunes were of for combat, or even that it was the white robe that candidates for the elections had mandatory in Rome; it was then toga candida. In the one case, as in the other, Herod wanted to demonstrate that they regarded him as a military leader, or the applicant for a function. The allusion is clear and reinforces our thesis, namely that persecuted Jesus as a rebel, as pretender to the throne, as a guerrilla leader fallen below by vital necessity, banditry, but in no case as a blasphemer. The process is a process of Jesus partly political and partly common law, without more, but both poles could not be separated. And this will prove it now analyzing the indictment.

The indictment of Jesus I love the curse! That falls therefore it on SALMOS, 1Ü9, 17 The various disturbances provoked by the messianic and fundamentalist activity of Jesus, which we shall call the "Great Revolution", considering further importance in the history of the world, and that would not end until the end of the age of Pisces, lasted about four years, at most! To achieve evolve freely, followed by a mass of several thousand people, his supporters armed, accompanied by their wives and children, as was the custom throughout the Middle East, and living without work because, having left his usual life , they had become gradually people outside the law (bar Jonah, in Akkadian) and necessarily what they caught in their path, good or feed the poor (Mark, 6, 36), it was necessary that Jesus benefited from fear or tacit complicity of sedentary and not "committed" in all populations. And the same in Jerusalem, and the following passage from the canonical Gospels proves indisputably:

"That day came some Pharisees to say," Go away, leave here, for Herod wants to kill you ... ' "(Luke 13:31.) 25 And if we refer to John (7, 30 and 7, 44) we see how scurry Temple militants not to be arrested, and Sanedritas content, good-natured, to his explanation. It is easy to understand that those passages were conceived from start to finish by anonymous scribes of the fourth century for the sole purpose of trying to provide an explanation to this astonishing and permanent impunity. Because, at that time, it was unthinkable that militiamen or (25 This is Herod Antipas, obviously) some dark guards could freely assess an order from the legitimate authority to decide whether it should be executed or not for them. And on the other hand, for twenty centuries, disobedience of the soldier will be punished with death, in all armies of the world. Thus. Jesus enjoyed long discrete benevolence of some and the prudent neutrality and hostile indifference of others. But one day Rome finally exhausted his patience and decided to end it, and then had to be imperative that official Judaism take sides. It is likely that Pilate decided to take hostages, or even strike at the Jewish community indiscriminately, as believed, rightly accomplice of Jesus. And as the Sanhedrin, also he played you choose. A phrase of the Gospels confirms this: "One. Caiaphas, who was high priest that year, said to them: You know nothing! Do you not understand that it is better for all that one man die for the people, shall not perish and the entire Jewish nation '! "(John 11,50.) Thus, the activity of Jesus and his band of zealots had finished by putting the entire Jewish nation in danger of perishing. This will not surprise anyone if the accounts of Flavius ??Josephus in which he sees the Romans deported and sold as slaves to the entire population of some villages, guilty of having supported the Jewish resistance are remembered. However, a point that absolves Caiaphas the high priest of all egotistical calculation is that the Gospel of John, in that passage, we specify that one uttered those words, not by itself, but in a true prophetic delirium, i.e., low divine inspiration, which recognizes the gospel itself in that circumstance. It is, probably, that phrase, so clear, so simple, where Paul, the "visionary", extrapolating the idea that Jesus died for spiritual salvation (and no longer stock) of all nations (and no longer Israel only). Therefore, it was to flatter the imperial power, Rome and Constantine so that anonymous scribes of the fourth century, who were already anti-Semitic, endeavored to present Jews as if they had been fierce with Jesus, to lose, and striving to acquit Pilate, when surely it should be just the opposite.

Six years Juan Solana, a Catholic priest ago, bought a property in the ancient town of Mandala and was forced to make exploratory excavations under Israeli law. By chance he found the remains of a 1st century AD synagogue.

"Historians believe that Jesus may have once walked the cobbled streets," says he was released WDAM.com "This may have been the home of one of the most important figures of the Bible, Mary Magdalene. The first witness of the resurrection registered. "This is a sacred place. I'm sure," said Father Juan Solana. Mysterious by the structure found in the Sea of Galilee perplexes scientists New Mosaics Add to the intrigue of the story of Israel created, Synagogue.

Lifestyles of the rich and famous times in the Bible Smithsonian.com says that for centuries Mary Magdalene was "the most obsessively revered saints, this woman became the incarnation of Christian devotion, which was defined as repentance." The article says that is almost certainly false security that she was a repentant prostitute. St. Gregory the Great in the 6th century D.C. was the first to call Mary Magdalene, sinner, per the Book of All Saints by Robert Ellsberg. She is identified in the Gospels of Mark and Luke as "a woman of whom seven demons had gone out." Ellsberg writes that the statement about demons can be interpreted in various ways. From this perspective, the mother of the seven thunders refers to Mary his mother.

Juan Solana, a Catholic priest, bought a property in the ancient town of Magdala and was forced to make exploratory excavations under Israeli law.

By chance he found the remains of a 1st century AD synagogue.

"Historians believe that Jesus may have once walked the cobbled streets," says he was released WDAM.com "This may have been the home of one of the most important figures of the Bible, Mary Magdalene. The first witness of the resurrection registered. Man, by the structure found in the Sea of Galilee perplexes scientists New Mosaics Add to the intrigue of the story of Israel Synagogue "This is a sacred place. I'm sure ', Father Juan Solana said". Lifestyles of the rich and famous in the Bible.

Smithsonian.com says that for centuries Mary Magdalene was "the most obsessively revered saints, this woman became the incarnation of Christian devotion, which was defined as repentance." The article says that is almost certainly false security that she was a repentant prostitute. St. Gregory the Great in the 6th century D.C. was the first to call Mary Magdalene, sinner, per the Book of All Saints by Robert Ellsberg.

disciples with the news: "I have seen the Lord!" And she told them that he had said these things to her.

Because Mary Magdalene was the first to tell the apostles about the resurrection of Jesus, who has been called the Apostle to the Apostles. The Do Chemical analysis of bones ancient stone box rekindles debate on the alleged family of Jesus.

Solana Magdalena bought the land to build a Christian retreat. Dates of the synagogue of the first century A.D. The New Testament says that Jesus preached in the synagogues of Galilee, and this is the only dating from the life of Jesus that has been excavated in the area.

"Not for me. This is for millions of people who will come to see this is going to enjoy this as I did and is expected to be able to discover our common roots"

The synagogue is adorned with frescoes and mosaic floors. It has an altar, called in Hebrew Binah in the middle. People call this the Magdala stone. It has in it a rare menorah carved in stone.

The synagogue is adorned with frescoes and mosaic floors. It has an altar, called in Hebrew Binah in the middle. People call this the Magdala stone. It has in it a rare menorah carved in stone.

The city of Magdala purification bath, which is still functional, is located at the bottom of the visible steps in the center of the photo. (Wiki Commons media)

Archaeologists say the synagogue in Magdala is one of the most important discoveries in Israel in 50 years. No also they found a container dating back some 2,000 years that Jesus may have washed their hands before preaching or praying in the synagogue.

Marcela Zapata, an archaeologist, describes the purification baths, which still work. "It is the purest water in all Israel Today if I ask some volunteers to take all the water and to clean the floor and steps in half an hour water begins to leave again.

Again, highlights the continued creation of myths about the life of a citizen of Israel Jesus who lived his life as a Jewish leader, the Menorah demonstrates, this is not Christian symbols.

Jesus did not found any Christianity, Later studies About His Life mixed the history of the Galilean Zealots and confusing story invented by Paul of Jesus life as Christians.

Mother of Jesus visit the tomb early Sunday 15 April- resurrection

Maria Magdalena - Was a creation or a detour to avoid revealing the true story of Jesus mother?

The creation of the story of Mary Magdalene used in the writings, was a creation of the centuries IV and V to hide the true story of Jesus and highlight other personalities to call the attention far form Jesus. Neo Testamentary writings identify this woman as Mary the mother of Jesus.

Example of contradictions call the attention to what confuse an investigator.

To draw attention against of Jews as responsible for the Jesus death. The main object was to free Roman blame. Another point in the same direction is, when Pilate says to the Jews "I give him that right, I see no fault in him" Unison all Jews who were present, a crowd answer at same time; "That the fault of this just fall on us and all our descendants", a crowd they shouted a statement at the same time without having agreed each other.

Another important fact that draws attention is when he enters the stage "Bar -Abbas- The Jews claimed that to free Bar- Abbas. Bar Means Son, Abbas Father. It was the Jews who asked for the release of Son of the Father's Jesus" As it was reached to weave this story for posterity. In the accusation act Pilates and the slandering accuse Jesus as Bar

Abas. If Pilates obey the Jewish claim, Jesus must be free. In addition to the laws no more than one man could be crucified at the same time. Talmud and Jewish law, the declarations in the scriptures is a fraud.

The charge of all the Jews of killing Jesus was the excuse then exterminate six million Jews. The anti-Semitic persecution during the Middle Ages, when Hitler and Franco used the extermination of Jews under the approval of Pope Pius XII and communism of the Jesuits. Hitler was brought to power by these created characters to erase the Jewish people from the earth.

Crucifixion weekend

Mary Mother of Jesus:

The first day Maria came from

Early morning, when it was still dark, and saw the stone taken away. Mary stood near the monument as he cried, bowed to the monument, and saw two angels in white sitting one at the head and one at the feet where the body had been, they said, why are you crying woman? She told them "because they have taken my Lord and do not know where they have put. Saying this apart turned back and saw Jesus standing there, but did not recognize it was Jesus.

Jesus said unto him Woman Why are you crying? Who are you looking for? "She thought it was the gardener said, Sir if you have carried him away where you have put him and I'll take?

Jesus said to him:

"Maria" She turned and said to him in Hebrew, Rabboni, "which means Teacher. Jesus answered him. "Do not touch me because I have not yet ascended to the Father" (John 20: 1 to 17)

Gospel of the Twelve Apostles:

Origen regarded as one of the oldest, prior to Lucas.

She opened her eyes because he had pulled down.

She seeing said cheerfully; Teacher, my lord! My son, you have truly risen!

Note:

Everyone knew of his powers, but they doubted and he had promised to resurrect. No mention of his rise in these writings, because it was visible to all. I wanted to catch him and kiss him, but he stopped her saying, "Mother, do not touch me, wait a minute ... is not possible that anything carnal touch me until fence to heaven. However, this is the body that pass nine months in your womb. You know these things or my Mother, know I'm who you fed. Do not hesitate mother that I am your son. I'm the one who left you in the hands of John when he was hanging on the cross. Mother now hasten to tell my brothers. (Gospel of the twelve Apostol's)

Gospel of Bartholomew;

In the tomb, Jesus shout "Marikha, Marima! Thiat! - Mother of the Son of God Almighty ... My Lord and my son.

El Salvador told Cheers to you who have taken the life of the world!

Health to you, my mother, my holy ark, my city and my resting place. Go with my brothers and tell them that I have risen from the dead ... (CF Gospel of Bartholomew, fragment)

Gospel of Gamaliel

Considered the most faithful to the truth.

Early Maria mother of Jesus, was beside the grave of her son. Which it is even more credible. Furthermore, to present a woman of questionable morals, who belonged to the family, as the first to be presented to the meeting with her dead son, leaving the mother alien to this pious duty. Per this gospel, Maria did not find the body of his son, but he argued with a stranger, she assumed it was the gardener, as in the aforementioned texts canons. "Lord this is what saddens me, because in this tomb I have not found the body of my son, beloved, to mourn over him, which would have consoled my sadness ... Now if you are the guardian of this garden, I conjure you tell already shed enough tears so far, look at my face, my Mother to convince you that I am your son. Then she said, "You have risen oh my lord and my son" (cf. Gospel of Gamaliel, extracts)

It is a demonstration outside the Gospels changed, these early Coptic Gospels tell us about the mother of Jesus.

The Origins of Christianity in Antioch

Paul Richard Quote

Summary

Reconstruction Movement of Jesus and the Christian churches in the city of Antioch in Syria and surroundings. We stress the importance of this city, third after Rome and Alexandria, where the meeting between the Semitic and Greek cultures is given. We distinguish three Christian generations.

The first (A.D. 40-70 years) is reconstructed from Galatians 2 and Acts 6-15.

Here we insist on the conflict of Antioch and its impact on the identity of the Church. The second (8090 years) is the generation marked by the Gospel of Matthew, which includes more than 50 years of tradition Galilee Jesus Movement and proposes an ecclesiology founded on the historical memory of Jesus, alternative to the synagogue Judaism rabbinical post 70AD the third generation (years 100-130) is represented by Ignatius of Antioch and the Did ache, which still survives the charismatic and prophetic tradition of previous generations.

There are several occasions on which biblical or rabbinical regulations require immersion of the whole body, referred to as *tevilah*. Depending on the circumstances, such ritual bathing might require immersion in "living water" - either by using a natural stream or by using a mikveh (a specially constructed ritual bath, connected directly to a natural source of water, such as a spring).

This article discusses the requirements of immersion in Rabbinic Judaism and its descendants. Some other branches of Judaism, such as Falashas Judaism, have substantially different practices including the requirement of an actual spring or stream. Vault and a camera with three bread ovens.

Inside a room is intact roof with a bathtub - a very rare commoners of the time could not afford luxury. read more: Jewish newspaper" Haaretz

http://www.haaretz.com/jewish/archaeology/1.73 0 486

A replica disc Emperor Theodosius, who condemned the destruction of the Jewish houses of worship, to the total indifference of Bishop Ambrose. Wikimedia Commons

read more: http://www.haaretz.com/jewish/thisday-in-jewish-history/1.734468

"Prof. James Tabor of the University of North Carolina at Charlotte told Haaretz." Haaretz is a popular diary of Israel. Caiaphas was the son of Anna's, who had six children, where high priests were passed in the show about 60 years putting their children in one and then the other, and his son-in-law, Caiaphas along the way. So maybe these are the homes of the extended priestly dynasty"

Tabor said

The high priests at that time were notorious, with a reputation for being corrupt, brutal and greedy. We learn not only from the Gospel accounts, but from texts of the Talmud tells how the high priest used to beat people with sticks (Pesachim 57a, Pp.284-285). Josephus gives us an equally grim account of the pontiff a 'hoarder of money "(Josephus - Antiquities of the Jews 20.9.2-4:" But as the high priest Ananias, which increased in glory every day, and this greatly, he had obtained the favor and esteem of the citizens of a form of signal, because it was a great hoarder of money").

read
http://www.haaretz.com/jewish/archaeology/1.730

Archaeologists digging in the heart of ancient Jerusalem have begun to discover the neighborhood that housed 2,000 years ago, the elite - most likely the priestly ruling class.

One house had its own cistern, a mikveh (a pool of Jewish ritual bath), a roof barrel vault and a camera with three bread ovens.

Inside a room is intact roof with a bathtub - a very rare commoners of the time could not afford luxury.

206-THE SECRETS OF GOLGOTHA

Related version of Jesus life. Jesus Bar-Juda. How he was censured Tacitus, Suetonius and Josephus, to better support the legend of an incarnate Good.

Jesus-Barabbas. Impossibility of penal substitution in Jerusalem at that time. By what this imaginary character, intended to mask the Zealot activity of Jesus.

The secret of Simon of Cyrene. A discreet controversy among exegetes of the first centuries. What masking that discussion.

Jesus evasion. Sacked six weeks before Easter, absconded with the agreement Tacit Pilate, revolting Samaria. It is caught again in Lydda and returned to Jerusalem, where he is crucified.

Two falls in sick mysterious misfortune. Pilate is denounced by the Sadducees for having Jesus allowed evasion and therefore the revolution of the Samaritans. Is exiled to Vienne, where he died. In turn, Herod Antipas is exiled to Vienne. Real motives.

When Jesus died? Why they are erroneous advanced by the official exegete's data.

How to calculate the exact day and year of Jesus' death.

The mystery of the tomb. Did Jesus have the privilege of having a ritual grave, or was released to the fossa infamy, like all death row?

On the cremation of the body of Jesus in Makron, Samaria, August 1- 362, by order of the Emperor Julian. Impossibility in question from John the Baptist.

The resurrected on Good Friday. Inability to admit that story. It was about Zealots fighters hidden in the ritual cemetery on the Mount of Olives.

The shadow of Tiberius. Why the emperor thought make Jesus Tetrarch, or even a king of Israel. Jesus was a pawn in their strategy against the Parthians.

Mary, mother of Jesus. His genealogy. His doubts regarding the divinity of his son.

They raised the creation of the imaginary character of Mary of Magdala. also, he died in Jerusalem.

Large families: Hasmonean, Davidic, Herodian, the throne of Israel dispute. The stepsister Mary the mother of Jesus is none other than Mariamne II, aka Cleopatra Jerusalem, ninth wife of Herod the Great. Their plots and final.

"After the Ascension of Jesus, Judas, also called Thomas, sent to Abgar, king of Edessa, to Apostle Thaddeus, one of the seventy disciples ... ". (Eusebius, Ecclesiastical History, XXX, xx,

Jerusalem at that time. By what this imaginary character, intended to mask the Zealot activity of Jesus-Barabbas. Impossibility of penal substitution in Jesus was created.

The crime of the Temple. The road from Jericho to Jerusalem. The attack from the merchants and pilgrims. The makeup of words in the original stories.

The truth about the Passion. Impossibility of the farce of ridicule, contrary to Roman law, and its explanation; the facts on which further embroidered.

Real motives.

Declaration

As we have seen, the final battle took place in the Temple of Jerusalem transformed into a fortress by the insurgents, and Jesus alluded to the death of Zechariah, if we give credit to Matthew text.

"To me fall upon you all the righteous blood shed on earth, from the blood of righteous Abel to the blood of Zechariah son of Batrachian, 16 whom you murdered between the temple and the altar ... Verily I say unto you all this will come upon this generation ... ". (Cf. Matthew, (23, 35-36).

As we have seen, the final bout developed in the Temple of Jerusalem transformed into fortress by the insurgents, and Jesus alluded to the death of Zechariah, if we give credit to Matthew text:

"That upon you all the righteous blood shed on earth, from the blood of righteous Abel to the blood of Zechariah son of Barachiah, 16 whom you murdered between the Temple and the altar ... Truly I tell you, all this will come upon this generation ... ". (Cf. Matthew, 23, 35-36).

The sons of Aaron. Dual power between the Zealots. The truth about Zacharias.

The sons of David. Brothers and lieutenants of Jesus. Those who continued the fight against Rome, and those who defected.

Hezekiah-Har-Gamala. The ancestor of Jesus. Its operations against Syria. He is captured and crucify commanded by Herod the Great.

Juda-Har-Gamala. Son of Hezekiah, father of Jesus. What is known about him? His death in course of the Census Revolution, in year 6.

The James brothers. On the uncertainty regarding his position within the Davidic family. His death in Palestine and Jerusalem. The mystification of Santiago of Compostela.

Andrew, alias Eleazar, alias Lazarus. Brother of Simon Peter and, therefore, of Jesus.

Related to a " of resurrection theme."

The resurrection of Lazarus. What doubtful of such a miracle, ignored by Matthew, Mark, Luke and Paul. Possible explanation.

Judas bar- Judas, the twin brother of Jesus, alias Thomas, alias sends Lebeo, alias Thaddeus. He Attorney Cuspio Fado beheading.

Felipe. It is those who left the movement after the death of Jesus. What it ignores history.

Matthew. It is deserting the movement. Probably uncle of Jesus, perhaps father Juan de Gischala, another Zealot leader who will highlight during the siege of Jerusalem.

Bartholomew, alias Bar-Thalmai. Executed by order of Attorney Cuspio Fado, after his capture in Idumea.

Iochanan or John the Evangelist. Also, brother of Jesus. It was not ever in Rome,

but it was the religious leader of the Zealots. He died in Jerusalem while St. James the Less.

"Tongues of fire" of Pentecost. What was the "gift of tongues".

Psychiatric meaning of "glossolalia." What was the ritual of Tikun Chabouth?

Menahem, the "dildo" announced by Jesus. Grandson of Judas of Gamala, take Massada, then Jerusalem, is proclaimed king, he falls into a bloody dictatorship and finally is executed by the Israelites.

Simeon Bar-Cleophas. Descendant of David also, and crucified in Jerusalem after a new survey.

Simeon Bar-Kokheba. Called the "son of the star" supported by Rabbi Skiba, unleashes the great revolution of the year 135. At first get the victory, but then it crushed by the Roman legions, and will be responsible for the end of Jerusalem as a nation.

Mary, mother of Jesus. His genealogy. His doubts regarding the divinity of his son

They raised the creation of the imaginary character of Mary of Magdala also he died in

Jerusalem.

Large families: Hasmonean, Davidic, Herodian, the throne of Israel dispute. The stepsister Mary the mother of Jesus is none other than Mariamne II, aka Cleopatra

Jerusalem, ninth wife of Herod the Great. Their plots and final.

True Herod II: Lysanias, stepbrother of Salome II and her royal husband. What the imbroglio created by the copyist's monk's.

Daniel-Rops, Jesus in his son temps
You agree with this brief chronology of recent days lived by Jesus - Thursday, April 6: Dinner (at sunset), detention Olives; - Friday, April 7 (night) process, crucifixion, death; - Saturday, April 8: stay in the grave; - Sunday, April 9: Resurrection (at dawn).

We will now carefully study the claims of the Christian tradition, and to make their criticism. Those who drafted the centuries IV and V the Synoptic Gospels, the apocryphal John and did not have all the necessary elements to perform an unassailable work. Lacking communications, easy to consult libraries, epistolary relationships as comfortable as in our time, it was very difficult, if not impossible, to carry out a perfectly synchronized work. At that time, given our current methods of verification and control, he was not wanted fabulist who also were not even Jews.

His mistakes super abundantly prove. Customs and Jewish rites do not know everything, far from it. Here we will reproduce as relevant critical analysis of Auguste Hollard, in his Origenes des Fetes Chrétiennes: "The last meal that Jesus took with his disciples on Thursday, the eve of his death, left in the memory of these an indelible impression; It was then, for the last time his beloved Master gave thanks broke bread while, then distribute it as a symbol of union, and when filled the cup and blessed, before passing it to his disciples. "

There was nothing there that was not perfectly in accordance with Jewish custom, and to the

formulas of blessings, which read as follows: ". Blessed are you Lord our God, King of the Universe, which do produce bread to earth" and "Blessed are you, lord our God, King of the World, which has created the vineyard". It was during this meal when Jesus told his disciples: "I will not drink of the fruit of the vine until
I drink it new in the kingdom of God." (Mark, 14, 25.)

That's where it should take place the next meeting, hence there will be no opportunity or time to meet, because the Kingdom is coming. If Jesus has the feeling that, before inaugurate it, must pass through death, moreover, it is not sure. A few moments later, in the garden of Gethsemane, ask God to save this supreme test. "So, Jesus had been unable to think of founding, of that last meal and in commemoration of his death, an" institution of the Supper "that in any case the imminent prospect of a celestial appointment would have done well superfine.

The last supper of Jesus is not of any of the characters of the Passover meal, if not the final hymn (Mark, 14, 26 and Matthew, 26, 30) which, in any case, could designate the Hallel "Talmudic Discussion the liturgical use of Psalms 113-118 focuses on how the Psalms incorporate gratitude for past acts? God of salvation and confidence in the future redemption of Israel God "
But not in her or bitter herbs, and the four cups, even the Pascual lamb, which would have symbolized Christ better than any other food item, nor the unleavened bread, but ordinary bread (Arton in
Greek). '

Mark (14, 22-23) and Matthew (26, 26-27) we read: "While they ate. Jesus took bread, and blessing, broke it and gave it to them, saying, "Take, this is my body." Then taking a drink, after giving thanks, he gave it. " To view this meal a Passover meal, though it seems little- have to admit that cup of blessing which follows the distribution of bread was the third of the Jewish Passover ritual.

Lucas was more prescient and did start the meal (22, 17) with the blessing of the cup. He not puts him "as they were eating," which effectively disrupts the order of the food, and just the meal with a glass distribution, which could, in extreme cases, be very well the fourth ritual. " (Cf. Guignebert Jesus.) But we still expect other contradictions. How can we accept such absurdities from eyewitnesses, as John and Matthew, and that ignorance of the traditional, so punctilious, from pious Jews and Jewish ritual Luke and Mark '? For the Synoptics, that is, to Matthew, Mark and Luke.

Jesus celebrated the annual Passover before his execution, and handed them bread and wine, changed into mystical flesh and blood. For John, however, it was at the time the Passover, which rituals lambs were immolated in the Temple, whose blood would stain the altar (animals that parents were brought then home for preparing will consume by a family, according to a very specific ritual), at that moment it was when, for obvious esoteric symbolism, did expire Jesus on the cross.

Well, we have an obvious contradiction. For the Synoptics, the night before the day of execution on Golgotha, Jesus instituted the Supper, during his disciples. It so happened Thursday night, and as, per Jewish law, the day begins at sunset, was already the beginning of the 15th day of Nisan. During that day, it was when they had to sacrifice on the Temple Passover lambs. It was during the night that followed immediately when Jesus was arrested in the Garden of Olives, when he was tried and executed; therefore, it was the next day, i.e. Friday. Then he went into the tomb on Saturday and was resurrected on Sunday morning.

On the contrary, per John's account it was obviously a snack, a meal, and the episode of bread dipped in wine and offered to Judas it is proof of that. What does not say is that it was an institution of the Supper, a Passover meal or, in the ritual and Judaic sense of the term. Jesus' arrest also occurred the night of the 15th, but Thursday night of Nisan 14. The next morning, the Jews did not enter the Roman Praetorian for fear defiled and unable to eat at night the Passover lamb. (Cf. John 18:28.) And, therefore, it is the time when those lambs are sacrificed in the Temple, thousands, when Jesus expires on the cross.

We are at noon on Nisan 14. There are therefore two days apart with the Synoptic. And yet, these events, oh miracle! fall on the same days of the week: on Friday, the execution took place, and Sunday resurrection. The meaning of these special effects is unclear.
Because Friday is the day of Venus, aka Lucifer, and Jesus expires on the day of his Adversary. Hence the ban, for centuries, to celebrate the Eucharistic Supper dishes or cups that have copper in its composition, because this is the Venusian and Luciferian metal.
On Saturday, the Sabbath day of rest, is the day that happens in the silence of the tomb. And on Sunday, the day of sun, light, takes place at dawn resurrection. The events, as the count Matthew, Mark and Luke Synoptics, leading to impossible

anachronisms to admit, and show that the anonymous who wrote our Gospels in the IV and V centuries ignored the most elementary logic. If not, how admit that the first day of Passover, which must necessarily be devoted to rest, as inviolable as the Sabbath (Exodus, 12, 16), in a week that was a real spiritual "retreat" (pp. Cit., 12), they could occur mount the arrest of Jesus, deliberation of the accusers each other, and then with Pontius Pilate, the purchase of a painting by Joseph of Arimathea, and burial of Jesus?

In his Chronicle Pascóle (initium), the ancient author Apollinaire pointed out, rightly, that an execution in Jerusalem as sacred as the 15th of Nisan (April) day would have desecrated the Passover feast was prepared, and could have triggered an uprising over the Jewish masses. Rome, which was very prudent in these sensitive points, which had agreed to withdraw and hide the badges of his legions during his stay in Jerusalem, which had removed the shields of gold for the temple for being offered by uncircumcised, this Rome, which had shown many times its respect for the Jewish religion, it would not launch such a court challenge. Moreover, the Jews could hardly have been dispensed to attend the ordeal, they who (by the Gospels) Pilate had requested the arrest of Jesus. But the law says Passover explicitly: "[on that day] you do not occupy of any work." (Numbers 28:18.) During these holy days, Jerusalem was invaded by thousands of pilgrims.
Never the Roman Praetorian and the Jewish Sanhedrin could have come in that day to the trial of Jesus. When, some years later, Simon Peter also will be held during the Easter week (another uprising more), Herod Agrippa take care to

postpone his trial for "after Easter". (Acts of the Apostles, 12.4.)
In addition, the Synoptics themselves confirm to us that the arrest and subsequent trial could not take place these days: "They (the chief priests and scribes) said:" Do not be at the party, do not go to riot the people. "(Mark 14: 2 and Matthew, 26, 5.)
Other than that, the interrogation of Jesus during the Passover night was impossible legally, and we know how the Pharisees and the teachers of the law to these subtleties and girded those legal taboos. Indeed, in a city without night lighting, which, like all ancient cities, had a Draconian subterfuge (to alleviate the fire), it was physically impossible to gather immediately after the arrest of Jesus, and toward one of early morning, a whole Sanhedrin, consisting of seventy-two members, all elderly, the heads of the Cohanim, the scribes, the elders of the people and the numerous witnesses. Furthermore, per the law, the Sanhedrin, to judge in criminal matters, 50/0 could meet day and night ever "because the darkness clouding the judgment of man."

Moreover, in criminal matters, when the guilt of the accused was recognized, the verdict could not be taken until the next day. Therefore, per the law, "criminal prosecution could not ever start the day before the weekly Sabbath, or the eve of a religious festival" (cf. Michna, Sanhedrin IV, Babylonian Talmud, p.32). And there's more: it was not possible that 15 Nisan, analogous to compulsory rest day a Sabbath. Simon of Cyrene "came from the field," where have been working (Mark, 15, 21, and Numbers, 28, 18), not to be forced to help Jesus carry the cross, as this would have been a job. Finally, the output of Jesus, followed by his disciples after the Passover (or the "alleged"

Passover meal) food, described in Mark (14, 26), is incompatible with the formal requirement of Exodus (12, 22), which prohibits outright leave the house where the Passover meal takes place until the next morning: "Let none of you shall go out of the door of the house until the morning ..." (£; COAO, 12.22). on the streets, of
Jerusalem could not have, wandering around, but the Roman patrols, who were watching for a new uprising came to not disturb the party. And every Jew (easily recognizable by their typical customs) had been arrested on suspicion unerringly.
They are now a series of improbable things and apparent contradictions. The main reason that justified the arrest of Jesus was that claimed to be king. That would result in the inscription that Pilate himself drafted and sent nailing, by use of the time, above the patibular cross. And that was what the prosecutor reproached him during his interrogation, and that Jesus did not deny (Mark 15.2). Well, that is known as the crime of rebellion. And to be with Jesus, surrounded by his men, all armed with swords he had recommended them to be ready, if need be at the expense of selling their cloaks (Luke, 22, 36), Pilate orders a true armed expedition, comprising a cohort, i.e., six hundred veterans, elite soldiers commanded by a tribune, military judge with the rank of consul (John, 18, 3 and 12).
The contingent of armed Levites the Sanhedrin adds that little Roman army is not there but to manifest the loyalty of official Judaism. Everything seems therefore assume that, when Pilate who ordered that judicial expedition, he will take Jesus who once captured. Well then, that's nothing! Jesus, per the anonymous writers of our Gospels, will be brought before the Jewish religious authorities, and the whole process will focus, in fact, on a charge of blasphemy. At each end, he could have sustained

the hypothesis that was conducted before Herod Antipas, as this is the Tetrarch of Galilee and Perea, and to represent him their temporal power, legitimized by the agreement with Rome. Herod Antipas was precisely in Jerusalem at the time, in his palace, and Jesus, being a Galilean, depended on his authority. But our Gospels tell us that Jesus was led first: a) to "Caiaphas, the high priest" (Matthew 26.57); b) to "the high priest" (Mark 14.53); c) to "the high priest" (Luke, 22, 54); d) to "Anas, because he was father in law of Caiaphas, who was high priest that year ..." (John 18, 13). In the end, to whom Jesus appeared first? ¿First Anas or before Caiaphas? And Daniel-Rops notes with pregnancy:
"The annoying thing is that the text of the Fourth Gospel is very confusing at this point. We read that first led Jesus to Anna's, the father in law of Caiaphas, "high priest that year" (18, 13). Then comes a scene of interrogation, followed by the denial of Peter, who seems to be the same as the Synoptic located in Caiaphas; Then verse 24 states: "Anna's sent him bound to Caiaphas, the high priest." To achieve the logical sequence and both the agreement with the Synoptics, should we put verse 24 after verses 13 and 14, place which, incidentally, occupies an old Syriac manuscript and in Cyril of Alexandria. But then no word of what Anas said Jesus! "(Daniel-Rops, Jesus son temps, p. 496.) In fact, it is known, and involuntarily, a few pages later (p. 501) Daniel -Rops shows that during interrogation the pontiff said Israel could not lift in any way an accusation of blasphemy against Jesus. For this reason, we, for our part, in the event of the appearance of Jesus before the Sanhedrin see a sequence invented by anonymous scribes of the fourth century, who, being Greek and Semitic, tried

to liberate Rome from the responsibility of the Jesus' death.

Now, Christianity was not the official religion of the Roman Empire, and at all costs had to deal with punches to the imperial power. Instead, it is quite possible that Jesus was led first to present Tetrarch, since Herod represented the temporal Judaic power, while Roman Pilate represented the temporal, occupying and protecting power, and therefore superior. And, once again, the charge raised against Jesus is to be claimed king. We have the proof in this passage associated with the above activities of Jesus: "The same day came some Pharisees to say," Go away, leave here, for Herod wants to kill you. " He replied, "Go tell that fox ..." "(Luke, 13, 31.) Why did Herod Antipas, Tetrarch of Galilee and Perea, and at that time wanted to kill Jesus? Because the latter represented the Davidic and royal legitimacy, after his father Judas of Gamala, and declare it to be pretended king.

If not, what came that hatred of Tetrarch? What could give him some lessons of piety and collective moral taught the people? What could offend him the Gospel message intended? Anyway, the fact is that Jesus appeared before him after his arrest, and the story that make us about it contradicts the precedent:

"It Hearing of Galilee, Pilate asked whether the man was a Galilean, and having learned that he was from the jurisdiction of Herod, he sent it, who was also in Jerusalem in those days." When Herod saw Jesus, was very glad because had long wished to see him, because he had heard of him, and hoped to see him do some miracle. He gave a lot of questions, but Jesus said nothing. They were present the chief priests and scribes who accused him violently. Herod, with his escort, treated him with contempt and, after they had mocked him, having dressed in a shining robe, he returned to Pilate. On that day, Pilate and Herod became friends, for they were before enemies ... "(Luke, from 23.6 to 12.) Now, he says Daniel-Rops, a large part of the commentators estimate that the garment was a white robe analogous to the military tribunes were of for combat, or even that it was the white robe that candidates for the elections had mandatory in Rome; it was then toga candida. In the one case, as in the other, Herod wanted to demonstrate that they regarded him as a military leader, or the applicant for a function. The allusion is clear and reinforces our thesis, namely that persecuted Jesus as a rebel, as pretender to the throne, as a guerrilla leader fallen below by vital necessity, banditry, but in no case as a blasphemer. The process is a process of Jesus partly political and partly common law, without more, but both poles could not be separated. And this will prove it now analyzing the indictment.

The indictment of Jesus I love the curse! That falls therefore it on SALMOS, 1Ü9, 17 The various disturbances provoked by the messianic and fundamentalist activity of Jesus, which we shall call the "Great Revolution", considering further importance in the history of the world, and that would not end until the end of the age of Pisces, lasted about four years, at most! To achieve evolve freely, followed by a mass of several thousand people, his supporters armed, accompanied by their wives and children, as was the custom throughout the Middle East, and living without work because, having left his usual life , they had become gradually people outside the law (bar Jonah, in Akkadian) and necessarily what they caught in their path, good or feed the poor (Mark, 6, 36), it was

necessary that Jesus benefited from fear or tacit complicity of sedentary and not "committed" in all populations. And the same in Jerusalem, and the following passage from the canonical Gospels proves indisputably:
"That day came some Pharisees to say," Go away, leave here, for Herod wants to kill you ... ' "(Luke 13:31.) 25 And if we refer to John (7, 30 and 7, 44) we see how scurry Temple militants not to be arrested, and Sanedritas content, good-natured, to his explanation. It is easy to understand that those passages were conceived from start to finish by anonymous scribes of the fourth century for the sole purpose of trying to provide an explanation to this astonishing and permanent impunity. Because, at that time, it was unthinkable that militiamen or (25 This is Herod Antipas, obviously) some dark guards could freely assess an order from the legitimate authority to decide whether it should be executed or not for them. And on the other hand, for twenty centuries, disobedience of the soldier will be punished with death, in all armies of the world. Thus. Jesus enjoyed long discrete benevolence of some and the prudent neutrality and hostile indifference of others. But one day Rome finally exhausted his patience and decided to end it, and then had to be imperative that official Judaism take sides. It is likely that Pilate decided to take hostages, or even strike at the Jewish community indiscriminately, as believed, rightly accomplice of Jesus. And as the Sanhedrin, also he played you choose. A phrase of the Gospels confirms this: "One. Caiaphas, who was high priest that year, said to them: You know nothing! *Do you not understand that it is better for all that one man die for the people, shall not perish and the entire Jewish nation* '! "(John 11,50.) Thus, the activity of Jesus and his band of zealots had finished by putting the entire

Jewish nation in danger of perishing. This will not surprise anyone if the accounts of Flavius Josephus in which he sees the Romans deported and sold as slaves to the entire population of some villages, guilty of having supported the Jewish resistance are remembered. However, a point that absolves Caiaphas the high priest of all egotistical calculation is that the Gospel of John, in that passage, we specify that one uttered those words, not by itself, but in a true prophetic delirium, i.e., low divine inspiration, which recognizes the gospel itself in that circumstance. It is, probably, that phrase, so clear, so simple, where Paul, the "visionary", extrapolating the idea that Jesus died for spiritual salvation (and no longer stock) of all nations (and no longer Israel only). Therefore, it was to flatter the imperial power, Rome and Constantine so that anonymous scribes of the fourth century, who were already anti-Semitic, endeavored to present Jews as if they had been fierce with Jesus, to lose, and striving to acquit Pilate, when surely it should be just the opposite.

Additional notes

The present facts are bringing here summarizes exegete documentation to support not the Jesus historic events, just to show how His life was used to confuses the humanity.

Because the facts, and the time frame in Which the Control of the power do not let emerge the true historic facts for the future.

Daniel-Rops, Jesus in the book Jesus in his son temps

You agree with This brief chronology of recent days lived by Jesus - Thursday, April 6: Dinner (at sunset me from the cross.), detention Olives; - Friday, April 7 (nighttime from the cross.) process, crucifixion, death; - Saturday, April 8: stay in the grave; - Sunday, April 9: Resurrection. Daniel

Daniel Rops not mention in which year.

Abril 35

Dom	Lun	Mar	Mie	Jue	Vie	Sab
					1	2
3	4	5	6	7	8	9
10	11	12		14	15	16
17	18	19	20	21	22	23
24	25	26	27	28	29	30

calendario juliano

The year is 35 in the lunar perpetual calendar.

If we take the universal lunar calendar, we found difference between dates.

The pass over must be conducted in a full moon weekend as a Talmud and Jewish law. The only week that full moon happens was starting Monday April 11, of year 0035 during weekend of April 15, the only year this conditions are due.

Robert Ambelain investigations arrive to a different date and support with facts.

Jesus celebrated the annual Passover before his execution, and handed them bread and wine, changed into flesh and mystical blood. For John,

however, it was at the time the Passover, which rituals lambs were immolated in the Temple, whose blood would stain the altar (animals that parents were brought then home for preparing will consume a family, per a very specific ritual), at that moment it was when, for obvious esoteric symbolism, did expire Jesus on the cross.
Well, we have an obvious contradiction. For the Synoptics, the night before the day of execution on Golgotha, Jesus instituted the Supper, during his disciples. It so happened Thursday night, and as, per Jewish law, the day begins at sunset, was already the beginning of the 15th day of Nisan. During that day, it was when they had to sacrifice on the Temple Passover lambs. It was during the night that followed immediately when Jesus was arrested in the Garden of Olives, when he was tried and executed; therefore, it was the next day Friday. Then he went into the tomb on Saturday and was resurrected on Sunday morning.

On the contrary, per John's account it was obviously a snack, a meal, and the episode of bread dipped in wine and offered to Judas it is proof of that. What does not say is that it was an institution of the Supper, a Passover meal or, in the ritual and Judaic sense of the term. Jesus' arrest also occurred the night of the 15th, but Thursday night of Nisan 14. The next morning, the Jews did not enter the Roman Praetorian for fear defiled and unable to eat at night the Passover lamb. (Cf. John 18:28.) And, therefore, it is the time when those lambs are sacrificed in the Temple, thousands, when Jesus expires on the cross. We are at noon on Nisan 14. There are therefore two days apart with the Synoptic. And yet, these events, oh miracle! fall on the same days of the week: on Friday, the execution took place, and Sunday resurrection. The meaning of these special effects is unclear.

Jesus was in Jail with the disciples where he practices the las supper, unravel bread (Arton bread) with the family, event that use to introduce the communion in Christianity as a statement.

227- Short biography

Gaspar Pagan Chevere

Born in Barranquitas Puerto Rico, on January 7, 1945, son of Rosa Chévere and Arcilio Pagan.

Some titles published:

Enigmas of creation, the base that mature this book the origin of life, updating recent data, Gamala, land of Jesus- hard to believe religious statements on general findings and true witness-

Story Book "The Last Promise to my mother" "The Last Witness", History process on the promise I made to my mother before she died in 2008. Discover the murderers of my older brother in a US military base.

When trying to enter college in Puerto Rico in 68 it was hard for costs and the economy. Married in the same year, the obligations tied me to the daily routine. I managed to complete a course of mechanized accounting, with this base got work at the Medical Center of Puerto Rico and other companies. Coming to occupy supervisory positions, assistant manager at the International Airport in Carolina Puerto Rico. In 1985 he founded my own company, Pelulleras DC Corp. Acting as CEO for 30 years until my retirement.

My constant inner search led me to join AMORC- Ancient and Mystical Order of the Rose and Cross. It is a nonsectarian organization; their bases are widely known by search engines worldwide as initiatory institution. The great maturity in the knowledge I obtained in this way as a student to the present. My passion is researching topics about the history of Jesus, based on mystical experience. The search for details that have been changed by institutions through the accumulated history, to bring them to light, meaning that the reliability is not in historians.

Gaspar Pagan Chevere

Without being a mathematical or scientist delve into the secrets of cosmic creation.

Let's play dice with the universe (Book) is a collection of studies on the subject, my intuition let me travel through these hidden worlds and survive their discoveries. Commenting on Stephen Hawking, he projected a connoisseur of Christianity over a comment about Atheists, Christian concept to designate those who reject their teachings.

I communicate, that Atheists do not exist, there are only human beings who do not accept the way they are projected God, using mythical forms and idols to characterize it, creative mind to what we call God is the only way to theorize what you can call god, or consciousness of beings.

In the book Gamala Land of Jesus I inquire the details of the hidden story of his life. It was spread by Christianity in a dark age and used to guide the masses to its expansionist purposes, conquests of territories and the creation of an imperial nature elite, to the detriment of their adversaries in religion, this has not been corrected and now the same system still practiced.

Gaspar (Edwin) Pagan Year 2012

Second edition- August 2016

www.ingramcontent.com/pod-product-compliance
Lightning Source LLC
Chambersburg PA
CBHW070229190526
45169CB00001B/135
* 9 7 8 1 5 3 9 5 6 0 4 9 4 *